P9-DNN-455

Natural Disasters

The Earth, Its Wonders, Its Secrets

Natural Disasters

THE READER'S DIGEST ASSOCIATION, INC.
Pleasantville, New York/Montreal

NATURAL DISASTERS
Produced by Toucan Books Limited, London for
The Reader's Digest Association Limited, London
with Bradbury and Williams

Written by Dougal Dixon
Edited by Robert Sackville West and Helen Douglas-Cooper
Picture research by Marian Pullen

FOR THE READER'S DIGEST, UK
Series Editor Cortina Butler
Editorial Assistant Alison Candlin

READER'S DIGEST PROJECT STAFF, US
Editor Kathryn Bonomi
Art Editor Eleanor Kostyk
Production Supervisor Mike Gallo
Research Assistants Mary Jo McLean,
Valerie Sylvester

READER'S DIGEST ILLUSTRATED REFERENCE BOOKS
Editor-in-Chief Christopher Cavanaugh
Art Director Joan Mazzeo
Operations Manager William J. Cassidy

Library of Congress Cataloging in Publication Data
Natural disasters.
p. cm – (The earth, its wonders, its secrets)
Includes index.
ISBN 0-89577-915-3
1. Natural disasters. I. Reader's Digest Association. II. Series.
GB50 14.N37 1997
363.34–dc21 97-022869

Address any comments about Natural Disasters to
Reader's Digest, Editor-in-Chief, U.S. Illustrated Reference Books,
Reader's Digest Road, Pleasantville, NY 10570

To order additional copies of Natural Disasters,
call 1-800-846-2100

You can also visit us on the World Wide Web at:
www.readersdigest.com

Printed in the United States of America

Second printing, September, 2000

FRONT COVER *A tornado passing over Texas. Inset: A house explodes as it is engulfed by lava on Heimaey, off the coast of Iceland.*

PAGE 3 *Flooding in Khartoum.*

Contents

THE FORCES OF NATURE

While some of the worst disasters to befall people are the result of human activity, the most widespread destruction occurs as the outcome of natural processes that are way beyond the power of humans to contain or control.

In the early hours of May 22, 1915, near the town of Gretna Green in the Scottish borders, a crowded troop train collided head-on with a stationary local passenger train. A few minutes later, the wreckage was struck by a speeding express. Fire, triggered by exploding gas tanks used to illuminate the wooden carriages, engulfed the wreck and also two goods trains in nearby sidings. In this, the worst rail disaster in British history, 227 people died.

HUMAN DISASTER *The wreckage of three trains that collided near Gretna Green in 1915 continued to smoulder several hours later.*

Seventy-three years later, just before Christmas 1988, a terrorist bomb blew the nose off a Boeing 747 airliner almost directly above Gretna Green. The disabled aircraft glided into the town of Lockerbie 12 miles (20 km) away and exploded. The worst air disaster in British history claimed 270 lives. The stricken aeroplane missed the nuclear power station at Chapelcross by a mere 9 miles (15 km), so the disaster could have been much worse.

On April 15, 1912, the White Star liner *Titanic* was speeding across the North Atlantic in an attempt to win the Blue Riband prize for the fastest commercial crossing, when she scraped along the side of an iceberg. Her hull was ripped in several places, sending her to the ocean floor with 1500 passengers.

In December 1984, a chemical factory in the central Indian city of Bhopal developed a leak in a storage tank. The cloud of poisonous gas that was released killed 2000 people and subjected another 200 000 to long-term medical problems.

A clumsily handled safety test in a nuclear reactor near the small town of Chernobyl in the Ukraine on April 26, 1986, caused an unextinguishable carbon fire in the moderator, releasing a cloud of radioactivity that killed at least 250 people and contaminated thousands of square miles.

The events in this catalogue of disasters have an important feature in common – they were all caused by human actions. They would not have happened if the signalmen at Gretna Green had been paying proper attention; if the Lockerbie terrorists had not reacted to world events as they did; if the *Titanic* had been following a more southerly course; if the management at Bhopal had been more assiduous about safety inspections; if the

BATTLE FOR FOOD *Overgrazing on land in Kenya (right) has led to severe soil erosion. Top: Cleared land in Australia is ruined by the release of salt.*

technicians at Chernobyl had followed agreed procedures.

On another level, these events would not have happened if the Caledonian Railway had not had a junction with the Glasgow and South Western Railway at Gretna Green; if there had been no air traffic or sea traffic between Europe and North America; if Union Carbide had not built their chemical factory at Bhopal; or if the Soviet authorities had not chosen to use nuclear power as their energy source. All these disasters were the result of human activity, and so were avoidable in hindsight.

Then there are disasters that involve natural processes, but are triggered by human activities. History is full of tales of large-scale irrigation schemes that have unleashed floods when dams have burst, or contaminated soils with salts deposited from the water that was meant to bring life or, more unusually, altered the cycle of water evaporation and rainfall in an area so much that they have caused droughts elsewhere. Some regions of fertile soil have been so intensively farmed that the soil structure has broken down and the natural processes of erosion have carried the remains away, turning good farmland into desert. Infernos have raged through built-up areas with policies of fire-suppression, consuming leaf-litter and dead vegetation that has built up for decades – vegetation that would, under natural conditions, have been reduced by numerous small natural fires over the years. Coastal communities have collapsed into the sea when the land on which they were built has been washed away due to increased wave erosion, in turn brought about by the extraction of gravel from nearby beaches for building purposes, or by the construction of breakwaters to keep heavy seas out of sheltered moorings elsewhere. These disasters, too, might have been avoided if people had anticipated the effects of their activities.

Going Down *A landslide has carried away part of a house built above a beach. Bottom: Wildfire burns through Malibu, destroying 800 homes.*

Good and Bad Neighbours

Humans share the planet with millions of other living things. We have limited contact with most of them in our daily lives, but with some we have a closer, and often beneficial relationship. We breed domesticated animals and grow particular crops for food or clothing; we cultivate decorative plants and keep companion animals in our homes. In some cases, however, the living world works against us.

Blight *Potato plants show the effects of the fungus Phytophthora infestans.*

At the beginning of the 19th century, the population of rural Ireland was growing rapidly, with most people living on smallholdings on which only potatoes could be grown. Then, in the 1840s, a fungus called *Phytophthora infestans* took a hold on the potato crops and destroyed them. The growing plants died in the fields. Storage clamps began to sink in the middle as the potatoes disintegrated into a foul-smelling slime. People starved. The population of Ireland, which in 50 years had grown from about 5.5 million to 8.3 million, began to fall again. By the early years of the 20th century, the population had stabilised at a mere 4.5 million. Many had died of starvation, or from the diseases that hunger brings, but the vast majority had emigrated to countries such as the United States of America, which promised a better life.

The devastating disease known as the bubonic plague can be transmitted by black rats that carry fleas infected with the bacterium *Pasteurella pestis* in their fur. In the 14th century the increasing trade between Europe and the Far East gave the rats a mobility they had not experienced before. Ships trading between the Black Sea and the Mediterranean ports brought the plague westwards to a population that had

Starvation *An Irish family is portrayed during the famine of 1847, with nothing left to eat.*

no immunity, and the result was what modern historians refer to as the Black Death. Estimates of the casualties vary, but it seems that something between one-third and one-half of Europe's population died. After the first carnage, the disease remained endemic for three centuries, flaring up again locally from time to time and causing, among other outbreaks, the Great Plague of London in 1665. Nowadays, world health authorities deal with bubonic plague by controlling rats and fleas, athough it does reappear now and again, especially in tropical areas.

Locusts have plagued humanity since biblical times. When conditions are right and there is plenty of food – food destined for human consumption – locusts can form vast swarms that migrate across continents, devouring their way through any vegetation that they meet. The approach of a locust swarm is a dramatic sight. At first, it is a smoky smudge on the horizon. Then, as it approaches, it spreads across the sky like a black cloud blotting out the sun. The whirring noise of millions upon millions of wings becomes deafening. Growing plants vanish in a matter of minutes as each individual insect consumes its own weight of food in a day. Swarms have been measured that tower $2^1/_2$ miles (4 km) into the air, and some cover a ground area of 385 sq miles (1000 km^2). Boughs have been broken by the weight of insects settling on them. In one attempt to spray and swat a swarm out of existence, in Jordan in the early years of this century, 2447 tons of locust corpses were recovered. Any farmer up against a swarm that size faces ruin.

On a more widespread scale, but involving a smaller organism, the incidence of malaria is increasing. A tiny organism, *Plasmodium*, of which there are four species living in the blood, is transferred from one victim to another by the bite of a mosquito, thereby spreading the disease. In the early part of the 1990s there was a dramatic increase in the occurrence of malaria, and by 1995 it was claiming something like 2 million lives per year. That year, 280 million people were infected, and this figure is increasing at a rate of 5 per cent per annum.

The impact of malaria on a new area can have the proportions of a disaster. In the late 1980s there was an outbreak of the disease in central Madagascar, an area that had hitherto been clear: 100 000 cases were reported and 20 000 perished. In some areas more than half of those infected died. Many man-made factors have been proposed to account for this great increase, including the spread of human populations into regions from which tropical rain forest had been cleared, the onset of global warming, which increased the natural range of the carrier mosquito, and the reliance on long-used medicaments that have allowed the parasite time to evolve resistant strains.

Beyond All Control

Some disasters are brought about by entirely natural forces – they are unpredicted, and unavoidable even if they had been predicted – and these are the disasters that most concern us here. We live on a dynamic planet that, during its 4500 million years of

Unwelcome Visitors *A swarm of locusts is so dense that it darkens the sky. Top: A mosquito gorges on human blood.*

SPOT THE JOIN *The mountains of north-west Norway were formed when two continental masses collided.*

evolution, has been changing constantly.

The Earth's crust is continually shifting around. At the beginning of the Triassic period some 250 million years ago, all the continental masses of the world were fused together as one enormous supercontinent that geologists now call Pangaea. The rest of the Earth's surface was covered by the biggest ocean that ever existed – the superocean of Panthalassa. Before this time, the individual continental masses had been on the move, opening and closing oceanic expanses between them, and during the Triassic period they just happened to have all collided together.

We can still see the effects of these collisions in the ancient mountain chains that formed along the joins. The Appalachians in North America, the Scottish Highlands and the mountains of Norway were formed when a forerunner of North America collided with the continent that was to become northern Europe. Likewise the Ural mountains and their northern extension, the islands of Novaya Zemlya, were thrust up as the continent of northern Europe fused with that of Asia. No sooner had this supercontinent formed than it began to tear itself apart again. The Atlantic Ocean ripped it in two, India and Madagascar pulled themselves out of south-eastern Africa into Antarctica, and Australia finally parted from Antarctica. All this was accompanied by widespread volcanic activity. The Lake District of northern England is made up of ancient volcanoes that erupted as the northern continents ground into one another. Half of India is covered by lava flows that erupted at the time when it broke away from Africa. Volcanic activity is a sign that the surface of the Earth is continually evolving.

This movement is going on at the present time. India, having pulled itself away from Antarctica, has now collided with Asia and is still pushing into it, forcing up the Himalayas along the join. Australia is still moving northwards, twisting up the crust around it and forming the loops of volcanic islands of the East Indies. Africa is still splitting apart, as shown by the complex series of cracks that run up the eastern side of the continent as the Great Rift Valley.

HOT SPOT *The port of Heimaey, Iceland, is threatened when the neighbouring volcano erupts.*

The westernmost sliver of North America is shearing northwards along the edge of the rest of the continent and will someday separate completely. These movements are too slow for us to observe directly, but from time to time we are able to see the great energies that create them, in the form of earthquakes – convulsive movements that split the very foundations of the continents and settle them into new positions.

UPHEAVAL *An earthquake at Kobe, Japan, leaves a tangled mass of roadway and vehicles in its wake.*

When these great Earth movements take place, they cause disturbances to the waters of the oceans. Volcanic explosions or earthquake vibrations can set up waves – tsunamis – in the ocean waters. These build up into mountainous breakers and plunge down onto unguarded shorelines.

On land, the constant pushing of one continent against another crumples up the rocks along the boundary, gradually pushing up the great mountain chains of the world. Even when it is an area of oceanic crust that is pushing under a slab of continent, the edge of that continent is rising as a mountain chain. The Andes, which run the length of the western margin of South America, are being pushed up in this way. During this process, rocks are piled up beyond the level at which they can remain stable – the slopes become steepened beyond the rocks' angle of repose. As a result, rocks sometimes tumble down again, turning a vast area of mountainside into a landslide.

Owing to their angle of elevation or their latitude, the coldest of these mountains gather snow on their flanks. This, also, is unstable, and from time to time the snowfields collapse, sliding and tumbling as avalanches into the valley bottoms.

Above the crust and the oceans of the

Blowing up a Storm *Hurricane-force winds rip through Western Samoa.*

Earth is an even more mobile layer. The outermost part of the Earth's structure is formed by a gaseous envelope, the atmosphere. This mixture of gases lies on the surface, held in place by gravity, and becomes increasingly thinner the farther it is from the Earth's surface, until it trails off into the near-vacuum of space. Heat from the Sun passes through the atmosphere and warms the Earth's solid and liquid surfaces. Some of this heat is reflected back into the atmosphere and sets up convection currents that establish vast circulation patterns across the face of the Earth – movements that show themselves locally as winds. At times, these convection currents can become so violent that the resulting winds move vegetation and erode the landforms that they blast through. Hurricanes blunder through coastal regions, while their small, vicious relatives, tornadoes, bite and jump here and there in inland areas.

The heat of the Sun together with currents within the atmosphere also power the cycle of water evaporation and rainfall at the Earth's surface. This constant process transfers water between the oceans, the atmosphere and the land, converting water into vapour and ice, and back again. Rainstorms and rivers are a direct result of this activity, and once in a while they, too, become very violent. Rivers fill and burst their banks, spreading floods over regions of the land that are otherwise dry. On the other hand, atmospheric conditions sometimes deny moisture to areas that normally receive rainfall. The resulting drought can kill back the natural vegetation and have a devastating impact on animal life. Very dry areas are also particularly prone to fire.

The Earth and the other planets and their moons in the Solar System share outer space with innumerable smaller particles of matter such as meteoroids and comets. These proceed along much less predictable paths and, from time to time, fall under the gravitational influence of the Sun or other planets that they pass close to. Occasionally, a large fragment is

Floating Village *Torrential rain causes floods that isolate a village in north-eastern France.*

caught by the Earth's gravitational field and drawn into it. A very large meteorite can punch right through the atmosphere and hit the Earth, leaving a vast crater, and even disturbed ecosystems.

The Human Perspective

These events come and go, adding to the surface features, gradually altering the landscapes, but causing nothing that would be regarded as lasting damage. It is only through the perspective of human experience that such occurrences are demonised and take on the mantle of disasters. Just as there are no such things as pests – only animals that get in the way of human society – and just as weeds are merely plants growing in the wrong place – from the human point of view – so natural disasters are geological, oceanographic or meteorological occurrences that happen to cause widespread disruption to humans.

The principal reason for this attitude is human self-awareness. When a lemming population expands beyond the capability of its environment to support it, it migrates towards areas where more food may exist, but a vast proportion perish on the way. The lemmings do not see this as a disaster, however. It is part of their natural cycle. Imagine if this were to happen to human beings. With their natural concern for their fellow members, they would regard deaths in such numbers as appalling, and the environmental pressures that brought about the situation would be seen as a disaster of the first order.

Another aspect that affects the human view of these events is one of time scale. The Earth has been shaped by 4500 million years of geological activity. But human civilisation has only been going for about one-millionth of that time. The individual human lifespan is less than a century. Each human being sees very little of these movements that build mountains and split continents. Yet these great forces are churning away beneath our feet all the time. When these forces manifest themselves in a sudden short-term burst, as in a volcanic eruption or an earthquake, the result seems to us an event that is completely out of the ordinary.

Since the beginnings of civilisation, humans have come to regard themselves as the masters of the world, and are both surprised and horrified when the Earth's natural processes erupt and a city, a port or farmland happens to be in the way. Once the dust has settled, the affected people have been rescued and the rubble cleared away, we are left little the wiser – and are just as surprised and horrified when it happens again.

LUCKY ESCAPE *A comet burns a trail through the night sky, within sight of Earth but not making contact – this time.*

1 Land and Sea on the Move

Tortured Earth *The forces that build continents are revealed in the fissured soil of Japan.*

'As old as the hills', 'as solid as a rock', are just two of the phrases we use to express the permanence of the world around us. Some of our highest mountains, however, are among the most recent features of the landscape. They have developed and grown because of movements within the Earth itself, movements that can crumble rocks and create new ones, and that can displace the very waters of the oceans themselves. Most of these movements are only apparent over the millennia, but sometimes they happen with great suddenness in a particular place, and when they do, their effect on the human beings who live there can be disastrous.

Molten Mass *Lava solidifies as new rock in Hawaii.*

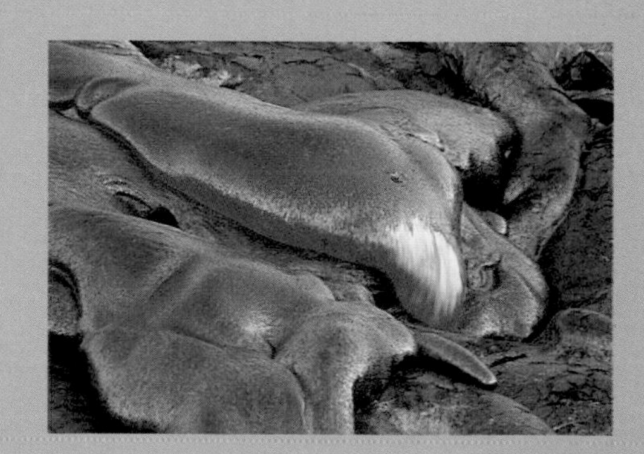

KILLER VOLCANOES

Deep below the Earth's crust, the temperatures and pressures are great enough to melt rock. When this molten rock blasts to the surface, the result is a volcano – and for the inhabitants of that place, it is as if Hell has been let loose.

The people of St Pierre, the main seaport of Martinique, were familiar with the rumblings and shakings that issued from the summit of Mont Pelée, towering 4583 ft (1397 m) above them, about 5 miles (8 km) away. However, the activity that began in April 1902 was rather more violent than usual. On May 2, after an intermittent series of explosions and gaseous eruptions, the crater lake emptied itself down the mountainside in a flow of hot black mud. The Governor of St Pierre assured everybody that there was nothing to fear — an election was imminent and he did not wish to see his voters fleeing to the countryside. In fact, the people from outlying areas were seeking refuge in the city.

Over the following week, 150 miles (240 km) away, the volcano La Soufrière, on the island of St Vincent, erupted with the loss of 1600 lives. The people of St Pierre believed this eruption would take the pressure off their own volcano but the tragedy of St Vincent would soon seem insignificant beside their own. On May 7, more ash erupted from Mont Pelée, but still the 29 000 inhabitants of St Pierre were assured that this was a passing phase. Then, at 7.52 the following morning – May 8 – they were engulfed by a *nuée ardente*, or incandescent cloud. Almost everyone was killed.

When magma – the hot liquid within the Earth's crust – is under pressure, it may contain a large quantity of gas in solution. If this pressure is released suddenly, there is an uprush of gases and expanding magma. Bursting through the flanks of a volcano as a white-hot mass of ash particles, and lubricated by expanding gas and steam, the magma pours down the slope at express-train speed, inundating and incinerating anything in its path. It follows gullies, stream beds, roads – any path of least resistance – at such speed that nothing can escape. Any living thing caught in its way dies instantly, its flesh and bones being reduced to cinders with the intensity of the heat. From a distance a nuée ardente has the appearance of a pillar of cloud, thick and hummocky like a cauliflower, with a roiling mass of incandescent heat at its base.

This was the horror that swept through the gardens, streets and squares of St Pierre in 1902. Trees, stores, houses and mansions were enveloped and burnt. Window glass melted, just split seconds before the frames disintegrated. Reinforced walls, more than a yard thick, were reduced to piles of rubble. Statues weighing tons were swept from their pedestals. The rum vats in the dockside warehouses exploded, pouring blazing rum

AFTERMATH *Mont Pelée broods over the silent ruins of St Pierre after the searing waves of its white-hot ash had extinguished life there in 1902.*

FIRE FOUNTAIN *Cascades of incandescent lava and glowing lava lakes (right) illuminate the night sky in Hawaii.*

Ash Cloud *A boiling mass of white-hot ash, fizzing with trapped gases, gushes into the sky under pressure from the Mount St Helens volcano.*

everywhere. The nuée ardente hurtled onwards and plunged into the harbour, hitting one ship broadside and capsizing it, and sweeping the masts and superstructure from others as they lay at anchor. Two ships were wrenched from their moorings and carried out to sea. Some of their crews survived to give eyewitness accounts of the disaster. On shore, the charred bones of 29 000 citizens lay beneath the settling mass of hot ash. At midday, the heat was still too intense to allow rescuers near the city. The ruins of St Pierre were still smoking three days later.

Two people in St Pierre survived. One, a prisoner, was deep in a cell in the town jail. The other was a shoemaker who hid beneath his work bench when the nuée struck. The prisoner was subsequently granted a pardon and went on to tour in a circus, advertising himself as the 'Famous Prisoner of St Pierre', the survivor of what proved to be the most terrible volcanic catastrophe of the 20th century.

The Restless Earth

The face of the Earth is constantly on the move, continually destroying and renewing itself. Imagine that this outer surface consists of panels, like the panels of a soccer ball; now imagine that each panel is continually growing from one seam, and is continually being destroyed at the other. These panels, or plates as they are known by geologists, are several miles thick and consist of the Earth's crust and the topmost layer of the stony portion of the Earth's interior, which is known as the mantle. The plates move about upon a soft, semiliquid layer within the mantle. It took geologists until the 1960s to discover this, because most of the evidence lies at the bottom of the ocean, which was inaccessible until then.

Along the floor of every ocean, there are ridges, which mark the limits of the Earth's plates – or 'plate boundaries' – at their growing edges, known by geologists as 'constructive plate margins'. The ridges at constructive plate margins have splits and rift valleys along (rather than across) them, peppered with underwater volcanoes and hot springs. The rocks of the ocean floor are quite recently formed, or 'fresh', close to the ridge, but become older farther away, as the matter welling up from deep within the Earth's interior moves away from the ridge and cools to form new plate material. As the new plates spread away from each other at the rate of up to 4 in (10 cm) a year, still newer matter rises up at the ridge, and so the process continues.

At its other edge, or 'destructive plate margin', the plate curls down and descends back into the mantle beneath the advancing edge of the next plate – a process marked by troughs deep in the ocean depths, such as those around the rim of the Pacific Ocean.

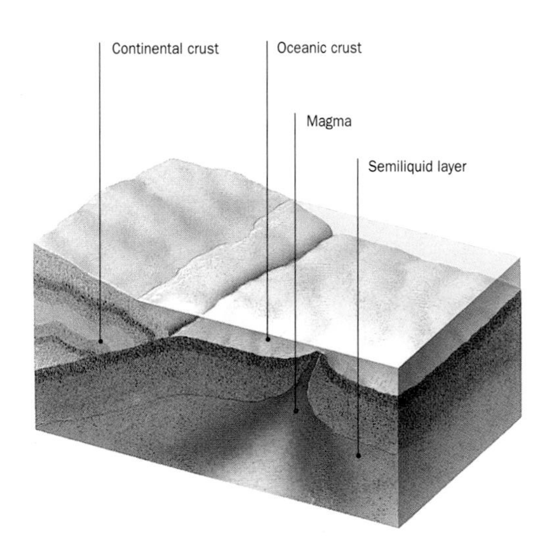

Crust in Turmoil *The great movements that trigger natural disasters all occur within the top few miles of the Earth's structure.*

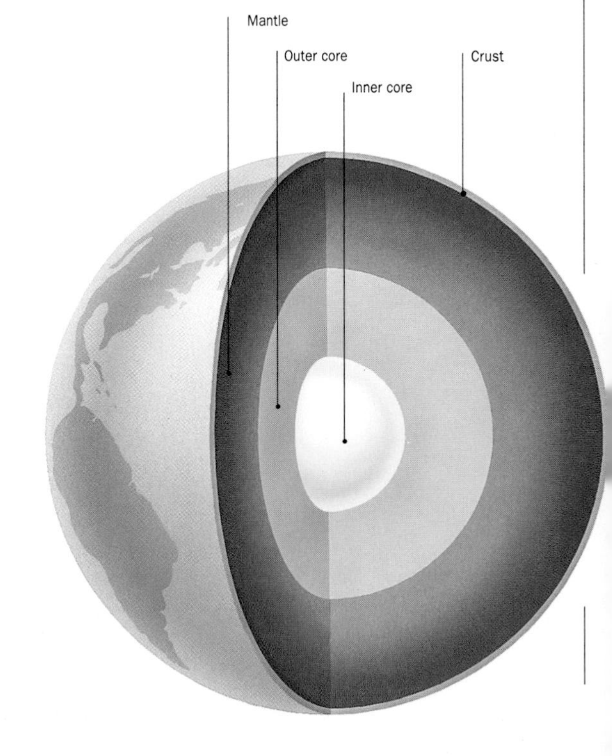

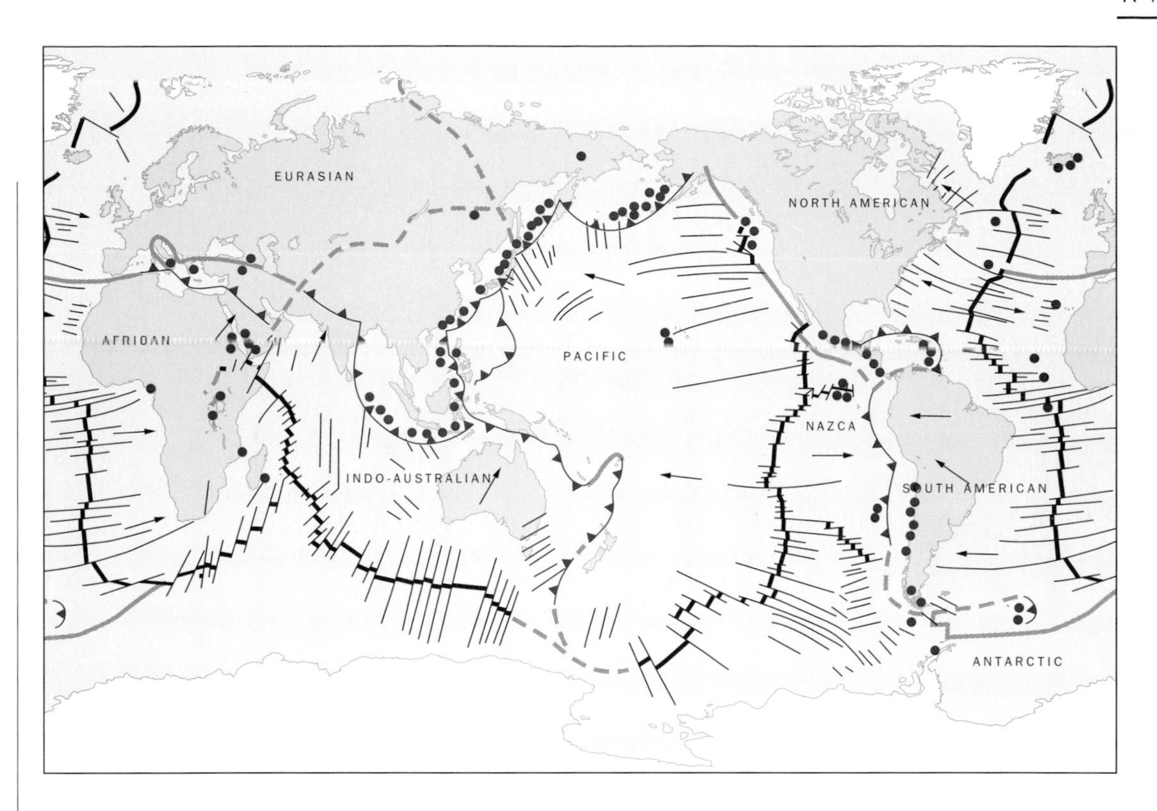

TECTONIC FEATURES

Plate boundaries
Transform fault
Constructive (ridge axis)
Destructive (subduction)
Undifferentiated
Uncertain and incipient
Direction of plate movement
Andesitic volcanoes
Basaltic volcanoes

RING OF FIRE *The distribution of volcanoes across the globe corresponds to the areas in which the plates are being created and destroyed.*

Those moving plates which have continents embedded in them carry the world's landmasses about as they go. The discovery of 'seafloor spreading' in the 1960s married up with an older concept of continental drift, and the two have since been combined to form the modern study of plate tectonics. Most volcanoes occur at the edges of the plates, where the activity is fiercest. Killer volcanoes, like Mont Pelée, lie at destructive plate margins. As one plate slides down beneath its neighbour, the two grind away at one another with unimaginable force, and the plate material melts with the heat of the friction.

THE WORKINGS OF A VOLCANO

The molten material, now called magma, rises through the plate above it and may eventually burst out at the surface to form a volcano. When this material erupts at the Earth's surface, it is known as lava and usually differs in composition from the parent magma, as the more volatile gas and liquid portions are given off into the air during the eruption, leaving only the thicker liquids and solids. The molten plate material at a destructive margin is rich in the mineral silica, which makes any resulting lava stiff and slow-moving. This type of lava – generally pale grey and cindery-looking – is known as andesite, after the Andes mountains, which are full of such volcanoes. If it erupts slowly, the lava forms hard angular fragments that roll downhill or off the front of a lava flow with a distinctive metallic clinking noise – rather like the sound of shunting railway wagons. Andesitic volcanoes occur in island arcs, such as Japan, the Aleutians, the sweep of the Caribbean islands and the East Indies, or in active mountain ranges, like the Andes and the Coast Ranges of western North America.

The volcanoes at constructive plate margins are quite different. The material that forms them is not recycled plate material but rises directly from the mantle, and this produces a magma with quite a different composition. Unlike the lava formed at a destructive plate margin, lava derived from the mantle at a constructive plate margin is low in silica and runny. Because it retains its heat, it is able to flow long distances before it solidifies. As a result, the basaltic volcano is very broad and low, and is often a relatively

continued on page 20

Ocean trench
Subduction zone
Magma
Ocean ridge

PLATE BOUNDARIES *At destructive plate margins (top), one plate is destroyed beneath another; at constructive margins (bottom), two plates are formed and then diverge.*

THE MECHANICS OF ERUPTION

Of the two major types of volcano, andesitic and basaltic, the typical andesitic volcano begins life when a mass of low-density magma forces its way to the surface. When the density of the rising magma is the same as that of the surrounding rocks, it gathers in a magma chamber. Any rise in pressure in this chamber may now push the magma upwards through cracks in the overlying rock. As the magma travelling up a crack approaches the surface, the pressure from the overlying rocks reduces; gases are released from the magma and expand so suddenly that an explosion rips open a funnel-shaped vent, called a diatreme, to the surface. The lava that blasts out of the vent then cools to form cinders, ash and dust – all referred to by the generic term 'tephra'. A ring of tephra collects around the vent and, as the eruption subsides, this blocks up the diatreme. Novarupta, in Alaska, is now at this stage of development. In time, subsequent eruptions blast new diatremes through the tephra plugging the vent, eventually building a volcanic cone. Occasional eruptions may blast an explosion crater, like a mini-diatreme, at the summit of the structure, and a fresh cone may then grow in the ensuing hollow. Vesuvius, in southern Italy, has reached this stage of its development.

The vent of an andesitic volcano may become stopped. If the pressure beneath the tephra and solidified lava plugging the vent reaches a certain level, it may be pushed up towards the surface, like the cork in a champagne bottle, to form a dome or a tall spine. A spine is a temporary feature, soon crumbling away to rubble.

Occasionally, the pressure within the volcano can build up to truly catastrophic

PROFILE OF TERROR *Andesitic volcanoes form steep-sided cones, such as White Island, New Zealand (below) and Cotopaxi in Ecuador (bottom).*

ANDESITIC VOLCANO

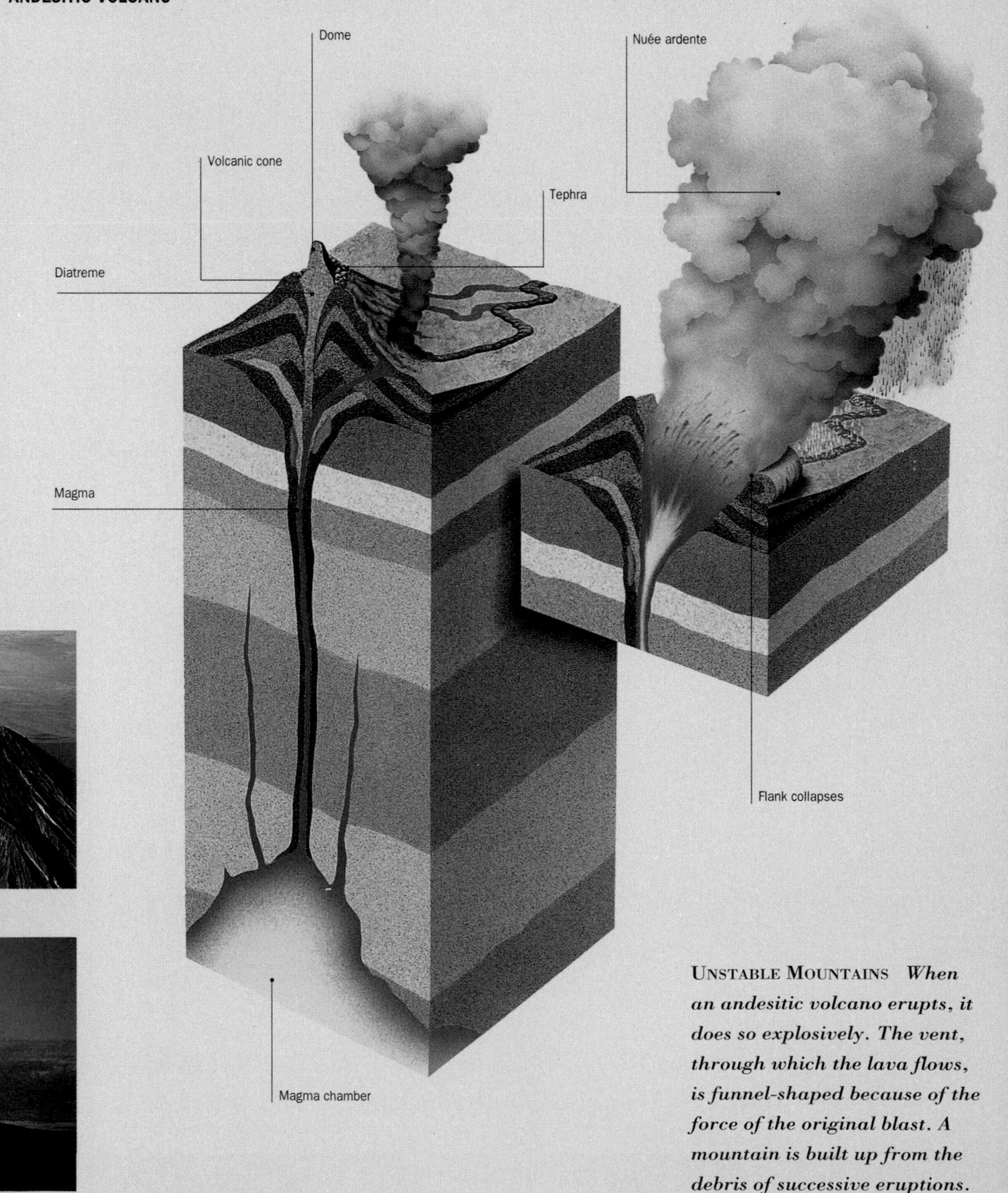

UNSTABLE MOUNTAINS *When an andesitic volcano erupts, it does so explosively. The vent, through which the lava flows, is funnel-shaped because of the force of the original blast. A mountain is built up from the debris of successive eruptions.*

LAYERS OF LIMESTONE *A hot spot sends chemical-rich water up as hot springs, laying down limestone formations.*

BASALTIC VOLCANO

Fire fountain
Magma
Dyke
Rivers of fire
Magma chamber

CRATER *Calderas are formed as a volcano's top collapses into a magma chamber.*

proportions, with the structure of the volcanic cone unable to contain it. The entire flank may collapse, releasing the pent-up forces as a vast explosion. A great volume of white-hot lava and ash, lubricated by expanding gases, pours like water down the remaining flanks of the volcano. This is known as a *nuée ardente* – a glowing cloud. Mount St Helens in 1980 is the best recorded example of such an eruption.

The magma that forms a basaltic volcano pushes its way from the magma chamber towards the surface through a crack. Many such cracks never reach the surface, and so the magma solidifies underground in vertical sheets, or dykes, of igneous rock. When the magma does force its way to the surface, it produces a fissure eruption as red-hot lava spouts out along the crack. Vast sheets of basalt rock, like those which cover half of India, are thought to have been produced by fissure eruptions in very ancient times. On a more local scale, such an eruption forms a shallow shield-shaped volcano. The eruption on the island of Heimaey, off Iceland, began with a fissure eruption in 1973.

LONG-TERM GROWTH *Basaltic volcanoes form broad shields covering vast areas. Eruption is through cracks and fissures. Lava lakes can fill the calderas that form during eruptions.*

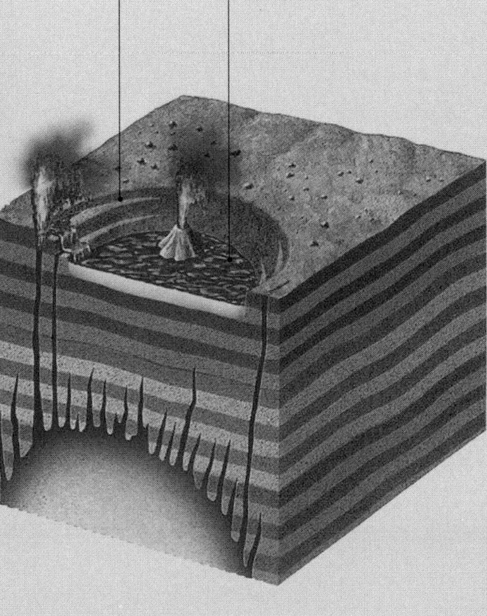

After a few hours, a fissure eruption usually settles down and the volcano continues to erupt from a single vent. This can produce an extraordinary fire fountain, as molten blobs of incandescent lava spout up into the air. The erupting lava is hot and runny, forming rivers of fire that run for long distances over the landscape.

RIVER OF FIRE *On the island of Heimaey, off Iceland, basaltic lava fountains out of the fissure to form glowing rivers that run for many miles.*

If the eruption continues, the magma chamber beneath will become depleted. The unsupported summit of the volcano may then subside into the chamber, forming a broad crater-like structure called a caldera. The original vent may continue to erupt, forming another, smaller volcano on the floor of the caldera, or the magma may take the easier course up through the cracks along which the caldera subsided. Fire fountains spray from these side cracks, and glowing molten rock cascades into the caldera. As it fills with molten lava, the caldera becomes a lava lake.

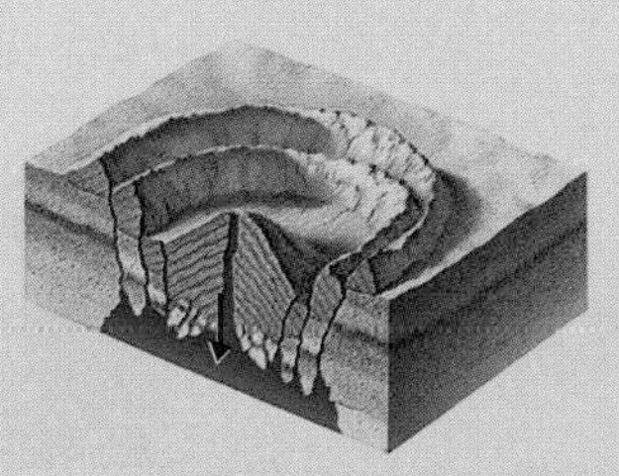

HOT SPOT *A chain of progressively older volcanic islands is formed as the plate moves over an active spot on the Earth's mantle.*

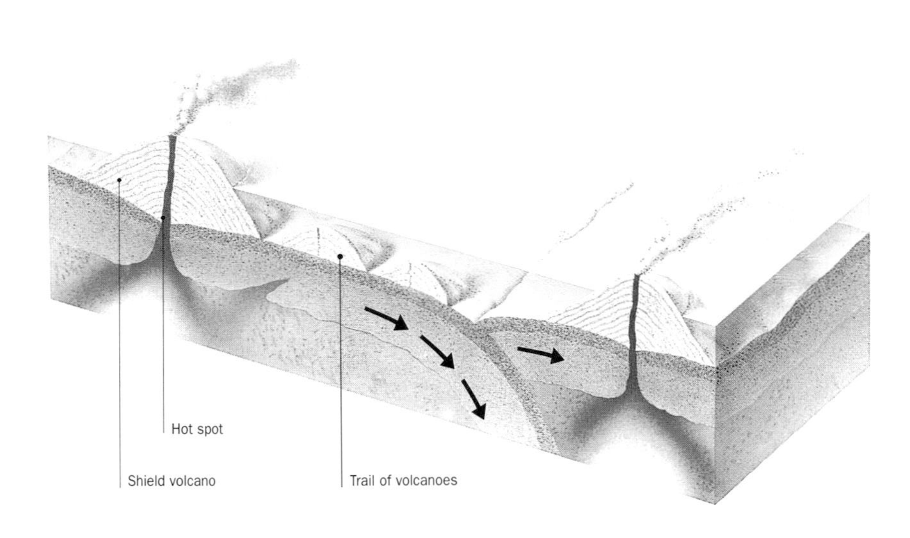

placid affair. Basalt lava, as it is known, is very dark in colour and its surface usually wrinkles up as it cools, giving it a ropy texture and its Hawaiian name *pahoehoe.* Basaltic volcanoes tend to occur along ocean ridges, but also in 'hot spots' where material rises straight from the mantle and punches through the overlying plate directly above it, far from any plate margin. Eventually the moving plate carries the volcano away from its original site above the stationary hot spot; and the volcano becomes extinct and begins to erode away. As it does so, a fresh volcano appears beside it on the adjacent part of the plate that is now over the hot spot. The result is one active volcano and a string of progressively older and eroded volcanoes, stretching away in the direction of the plate movement. Pacific island chains, such as Hawaii and the Galapagos Islands, are the result of hot-spot eruption. The Yellowstone region of hot springs and geysers in Wyoming, Montana and Idaho, is also attributed to a hot spot beneath the continent. Basaltic volcanoes do cause serious damage from time to time but are generally more predictable than andesitic volcanoes.

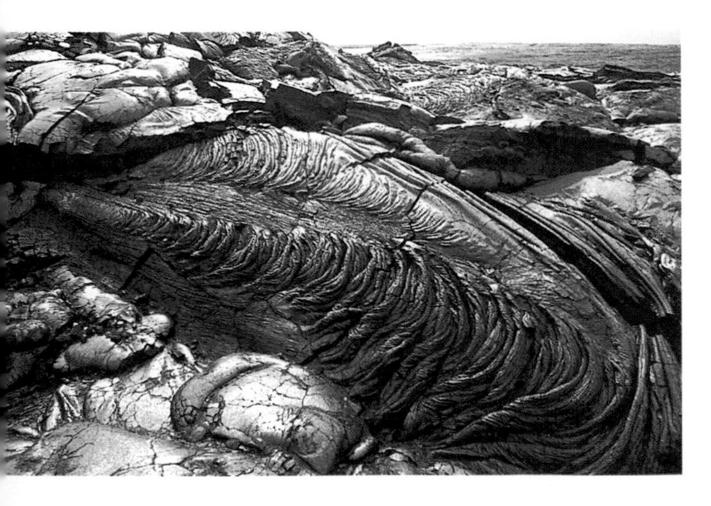

BASALTIC FLOW *As it solidifies, basaltic lava forms either a jagged blocky surface, known as* aa *(top), or a wrinkly surface called* pahoehoe *(bottom).*

A Basaltic Volcano Erupts

The kind of slow-motion disaster associated with a basaltic volcano is typified by the destruction of Kalapana, on the south coast of Hawaii, in 1990.

Kalapana was a quiet Hawaiian village subsisting on fishing and carob-growing until the mid 1960s, when its tranquillity and the beauty of its palm-fringed black beach – crumbling basaltic lava produces a distinctive black sand – came to the attention of the US mainland. Americans, attracted by the low property prices, bought plots of land and began to build their own houses, establishing the suburban community of Kalapana Gardens. The presence of the volcano Kilauea, 5 miles (8 km), away merely added to the romance.

On March 16, 1990, Kilauea began to erupt. A fire fountain fed a river of lava that flowed down towards the western side of the community as an incandescent red and white-hot torrent, blackening at the edges and in streaks, where the lava was beginning to solidify. As it spread out, it slowed, becoming a typical pahoehoe flow. In due course, the entire surface darkened and wrinkled. As the hot molten rock continued moving very slowly, it pulled this dark sticky crust apart, revealing the glowing interior momentarily, before the fresh surface cooled, darkened and wrinkled again. It crept over the ground, starting little fires in the undergrowth, but posing no real threat – children ran about and played ahead of its leisurely advance. For over a month, it spread seawards.

As one geologist on the spot remarked, it is only when the first road is covered or the first building catches fire that people acknowledge the seriousness of such a

Cooperation in Kalapana
Burning lava from Kilauea engulfs a house (right) in Kalapana. A church (below) was transported to safety.

situation. The evacuation of valuables and treasured possessions from Kalapana began more than a month after the start of the eruption, as the lava flow crept into gardens and sidled up against houses. Paintwork blistered and began to run; wood began to smoulder and finally burst into flames. The pace of the flow was so indolent that house-owners could walk away to safety, looking round and watching their homes destroyed. It was like a glacier, they said, slow and unstoppable – but a glacier with a temperature of 538°C (1000°F).

The lava advanced through Kalapana on a front several hundred yards wide, stretching across three to four streets and destroying a house a day on average. By May 1, when some 45 houses, representing about half of the community, had gone, the US Government declared a state of emergency. But the pace of destruction was increasing, and the original core of the town – two churches and a general store – was now threatened. On May 4, in a remarkable feat of community cooperation, contractor's equipment was brought in to lift the Catholic church from its foundations and move it out of harm's way. This was achieved only 45 minutes before the lava flow covered and destroyed the last road to safety. The last residents were evacuated on

continued on page 24

AT SEA *The basaltic volcanoes of Hawaii (top) and the andesitic volcanoes of Japan (above) are formed on the ocean floor, and their lava flows are uncontaminated by other minerals.*

ON LAND *Lava from volcanoes on continents, such as those in Zaire (right), absorbs the minerals of the continental crust through which it passes, and can produce strange chemical eruptions.*

May 5. The lava reached the sea on May 7, and on the same day the eruption began to die away. It started again three days later and, as a result, the general store, the remaining church and the famous black sand beach were engulfed.

About 140 houses had been destroyed in eight weeks. The damage was estimated at $60 million. Hundreds of lives were disrupted, but none lost.

Odd Volcanoes Out

Not all volcanoes fit neatly into the stereotype either of the andesitic volcano, like Mont Pelée, occurring at a destructive plate margin, or of the basaltic volcano, like Kilauea, occurring at a constructive plate margin. Sometimes processes take place within a basaltic volcano that change its nature dramatically. One of these, known as 'fractionation', can make the eruption of a basaltic volcano frighteningly like that of an andesitic eruption. If basaltic magma – low in silica by comparison with andesitic magma – cools and begins to solidify before reaching the surface, its component minerals crystallise out of the melt at different times, with the first solid crystals being those that contain the least silica. The melt that is left behind therefore becomes progressively richer in silica and begins to take on more of an andesitic composition. Should this melt and then erupt, the result is, in effect, an andesitic event in a basaltic volcano.

Hekla, one of Iceland's most famous volcanoes, was regarded by the early Norse settlers as one of the gates to Hell. In 1104 an andesitic-type eruption from Hekla destroyed a community of 20 to 30 farms and houses at distances of up to 45 miles (70 km) away – a most unusual event in an area of basaltic volcanoes.

Another volcanic exception to the rule takes place when the constructive plate margin lies in a continent. Such is the case with the volcanoes along the Great Rift Valley of East Africa, which runs from Mozambique northwards, through Tanzania, Uganda, Kenya and Ethiopia, continuing as the Red Sea and then as the valley of the River Jordan. The lava here starts off by rising from the mantle, but as it forces its way through the overlying mass of the African continent, it absorbs the minerals of the continental crust. As a result, the geology of these volcanoes is decidedly strange: some erupt minerals that form into lakes of washing soda;

Lake Natron *The chemical-rich lakes of Tanzania are the result of the mixture of minerals produced in the East African Rift Valley volcanoes.*

Fire and Ice *Although sited in a basaltic region, Iceland's most dangerous volcano, Hekla, can produce an andesitic-type eruption through the 'fractionation' process.*

THE BREAKUP OF AFRICA *Hot springs all along the East African Rift Valley illustrate the volcanic tensions as the continent pulls itself apart.*

and one even erupts a lava that turns into a kind of igneous limestone called carbonatite.

On the other side of the continent, an offshoot from the mid-Atlantic ocean ridge pushes eastwards into the heart of West Africa, coming ashore in Cameroon. Here the continental rocks are put under strain, producing a string of volcanoes which reaches inland, towards Nigeria and Chad, and continues to puzzle geologists. One, in particular, produced a strange and deadly eruption in 1986.

AN EXPLOSION OF LETHAL GAS

Lake Nyos, in Cameroon, is a crater lake, filling the crater of a volcano that last erupted several centuries ago. Deep below its 650 ft (200 m) of water, the blocked volcanic vents still send up carbon dioxide and other gases, which become dissolved in the water. Under the hot equatorial sun the surface waters of the lake are warm, but on the bottom the water is cold and dense and saturated with dissolved carbon dioxide. There is no circulation between the upper and lower waters.

During the evening of August 21 – for some reason still unknown – the dissolved gas in the lower waters of Lake Nyos surged to the surface. There appear to have been about 15 to 20 seconds of loud rumbling as the gas gushed upwards. What happened next is unclear. The carbon dioxide, a heavy but invisible gas, seems to have been flung out of the water into the air to a height of more than 300 ft (100 m) above the crater rim and to have fallen at some distance from the lake. Cows grazing nearby were unaffected. Once it hit the ground, however, the cloud of gas rolled down the mountainside at a speed of roughly 30 mph (50 km/h). In the village of Lower Nyos, less than a mile away, cooking fires were snuffed out and 1200 people, starved of oxygen by the descent of the carbon cloud, were suffocated to death. The cloud swept on down the valley, over fields and farms, killing people and livestock. In Subum, 6 miles (10 km) from the lake, patients on the first floor of the hospital and therefore above the level of the gas were unharmed, while those downstairs and in surrounding single-storey buildings perished. The gas cloud travelled 15 miles (25 km) before it dissipated, affecting an area of 11 sq miles (28 km^2), killing 1700 people and 3000 head of livestock. One or two people saw their families die but were inexplicably unaffected; others lost consciousness and awoke unharmed, sometimes 36 hours later.

In the aftermath of the gas eruption, it transpired that similar, less catastrophic incidents had previously occurred at the crater lakes of other volcanoes in the area. In this case, there seems to have been no volcanic eruption on the lake bed that could have disturbed the water. On the western crater wall, however, evidence has now been found of a recent landslide which, if it had taken place at the time of the disaster, might have produced a disturbance in the water. In any case, once the lower, gas-rich water had risen to a certain level in the lake, gas would have bubbled out as the pressure was released. This, in turn, would have set water circulating through the entire lake and all the gas would have broken the surface.

Scientists agree that gas is still seeping up into the cold waters of Lake Nyos – at a

LETHAL LAKE *Suffocating carbon dioxide gas, fizzing out of the water of Lake Nyos in Cameroon, killed 1700 people and their livestock in 1986.*

rate of about 175 million cu ft (5 million m^3) per year – but disagree about what should be done. Since 1990 a French team has been experimenting in Lake Nyos and nearby Lake Monoun (which has also released gas in this way). They have been pumping the cold water to the surface and letting it discharge its gases harmlessly, although other volcanologists insist that to do this without knowing a great deal more about the volcano may be rash.

Mud Slides

One of several deadly phenomena following in the wake of some eruptions is the *lahar*, or volcanic mud slide. Many geological events take their scientific names from the local language or dialect of the area where they are common, or where they were first studied, in this case, Java – mud slides being particularly prevalent on the steep-sided volcanoes of that island.

Much of the material erupted from an andesitic volcano comes in the form of loose ash, which is blasted from the crater and falls on the unstable flanks of the mountain. One of the commonest volcanic gases is water vapour, and volcanic eruptions are often accompanied by torrential downpours of hot rain, as water condenses from the gas clouds. The rain then mixes with the loose ash to form a slurry of mud that flows downhill, burying anything in its path. In Java, the Kelut volcano is a constant source of catastrophic lahars, and ever since the eruption of 1919 engineers have been digging tunnels through the mountain into the crater lake to drain off

River of Mud *When Nevado del Ruiz erupted, hot ash mixed with water from a glacier formed a mud flow (above) that engulfed towns and villages.*

Volcanic Mud *Loose volcanic ash, mixed with rain, produces devastating mud flows (below), especially in areas of tropical rainfall such as Indonesia.*

the water. These tunnels are continually being destroyed by successive eruptions, but they have gone some way to reduce the severity of subsequent lahars.

Perhaps the most dramatic lahar of recent years took place, not in Java but across the world in the Colombian Andes. Despite the fact that Nevado del Ruiz, an andesitic volcano 17 680 ft (5389 m) high, is almost on the Equator, its summit is covered in snow and contains a small glacier. In November 1984, the volcano began to erupt, exhaling steam and setting off a few small explosions which continued for about a year. On September 11, 1985, an eruption triggered a mud flow down the Azufrado valley on the north-eastern flank. Geologists investigating the occurrence were alarmed. In October they warned the Colombian Government that there was potential for a much larger mud flow, which could be disastrous in such a populated region. Little attention was paid to the warning; indeed, official proclamations, like those of St Pierre in 1902, assured the people of the area that they had nothing to fear.

At about 3 pm on November 13, 1985, the volcano began to throw out fountains of ash. The main eruption began around 9 pm, and for about three hours hot ashes

COLOMBIAN CATASTROPHE
Armero (left) was destroyed by the mud flow following the eruption of Nevado del Ruiz. Few people survived (below).

MAJOR VOLCANIC ERUPTIONS

Date	Place	Type	Death toll
AD79	*Vesuvius, Italy*	*Ash fall*	*20 000*
1631	*Vesuvius, Italy*	*Lava, mud and ash*	*4000*
1669	*Etna, Italy*	*Lava flow*	*20 000*
1772	*Papandayan, Java*	*Lahar*	*3000*
1783	*Laki, Iceland*	*Ash fall*	*12 000*
1802	*Mt. Unzen, Japan*	*Eruption, then tsunami*	*15 000*
1815	*Tambora, Sumbawa*	*Explosion, then tsunami*	*10 000 (directly)* *82 000 (famine)*
1822	*Galunggung, Java*	*Nuée ardente*	*4000*
1882	*Krakatau, Indonesia*	*Eruption, then tsunami*	*36 000*
1902	*Mont Pelée, Martinique*	*Nuée ardente*	*29 000*
1911	*Taal, Philippines*	*Nuée ardente*	*1335*
1919	*Kelud, Java*	*Lahar*	*5500*
1951	*Mount Lamington, Papua New Guinea*	*Nuée ardente*	*5000*
1980	*Mount St Helens, USA*	*Nuée ardente* *40 sq miles (100 km²) devastated*	*60*
1985	*Nevado del Ruiz, Colombia*	*Lahar*	*23 000*
1986	*Nyos, Cameroon*	*Gas eruption*	*1700*
1991	*Pinatubo, Philippines*	*Ash fall*	*900*

HOW SAFE IS A VOLCANO?

Until recently – at least, until the 1980s – volcanoes were often classified as 'active', 'dormant' or 'extinct'. A dormant volcano was expected to erupt again, while an extinct one was thought to be completely dead. However, geologists have been caught out so many times before by the eruption of supposedly extinct volcanoes that the classification has been abandoned. Nowadays a volcano is described as either 'active' or 'inactive'.

spewed from the vent, covering the snowfield and glacier. Below this hot blanket the snow and ice melted and began to pour from the peak down a river valley of the northern flank, picking up the freshly fallen ash as it went. Boulders and trees were ripped up as the torrent raced down the twisting canyon towards the town of Armero, which lay at the river mouth, 28 miles (45 km) from the crater. The tongue of black mud that gushed out of the canyon spread out in a blanket 13 ft (4 m) deep. Wave after wave of mud followed. It has been estimated that the flow of material in this mud slide was the equivalent of a fifth of the flow of the Amazon river. People and settlements in its path were swept away or buried. The death toll in Armero was 23 000 – 90 per cent of its population.

What Can Be Done?

The disaster at Armero led American volcanologists to set up a Volcano Crisis Assistance Team (VCAT), with an annual budget of $300 000 from the US Agency for International Development. VCAT consists of a group of volcanologists and a supply of equipment that can be deployed if any volcanic area begins to look threatening. It aims to produce a series of hazard assessment maps for every dangerous volcano, showing the likely extent of the damage in the event of an eruption.

Short-term prediction of this kind

Plume of Ash *When Mount Pinatubo in the Philippines erupted in 1991, accurate prediction led to successful evacuation operations.*

POZZUOLI, A PRECARIOUS PLACE TO LIVE

The Gulf of Pozzuoli, an inlet situated at the north-western corner of the Bay of Naples, may look like an ideal natural harbour, and it has been the site of a city since Roman times. Yet it is in fact a volcanic caldera, and volcanic activity has never been far away. The caldera itself was formed 34 000 years ago, and the last eruption took place in 1538 when a new volcanic cone – Monte Nuovo – was created. Early in the 19th century, geologists noticed that the columns of Roman temples in Pozzuoli showed traces of borings made by marine worms. These were about 10 ft (3 m) above the ground, which suggested that the ruins must have been submerged in the gulf and risen again at some time in the last 2000 years. For the next 150 years, the vertical movement of the area was monitored, and it was found to have been sinking at a rate of about half an inch (1.3 cm) per year as magma leaked from directly beneath. Then, in 1968, the movement changed direction and Pozzuoli began to rise again. The rising was irregular, taking place in several pulses with associated earth tremors, which severely damaged many of the town's buildings. The authorities decided to build a new town some distance away, and to evacuate most of Pozzuoli's 72 000 inhabitants. The area is still monitored for volcanic activity.

Bay of Naples *The cone of Mount Vesuvius is clearly visible in this satellite image.*

Shifting Land Surface *Former sea levels are visible on the ancient Roman columns in the Italian town of Pozzuoli.*

proved its worth when Mount Pinatubo erupted in the Philippines on April 22, 1991, after 600 years of tranquillity. There was a small explosion near the summit and new vents opened. The Philippine Institute of Volcanology and Seismology swiftly deployed its own monitoring technology, requested a team from the US Geological Survey, and evacuated 2000 people from the immediate vicinity. Instruments measured the earth tremors that originated within the mountain, the change in the angle of its slopes as magma was pushed up inside, and the change in composition of the gases being released. These readings were then analysed by computer. By June 5, the results indicated that a violent eruption was due within two weeks. By June 7, the readings predicted that the eruption would occur within 24 hours. In fact, the major phase began on June 12, but by then 64 000 local people had been evacuated. The death toll was around 900 – tragic, but less catastrophic than it could have been.

The geologists had managed to draw up a hazard map, which anticipated the possible areas of damage by ash falls, nuées ardentes and lahars. This map formed the basis for the evacuation programme.

A SUFFOCATING RAIN *Ash was the main product of the Pinatubo eruption, coating the landscape (below) and darkening the sky (bottom).*

ASH FALL *Houses and forests were choked by the heavy fall of ash from Mount Pinatubo, but there were fewer fatalities than anticipated.*

LIFE UNDER THE VOLCANO

Despite the constant threat of lava flows, nuées ardentes, gas eruptions and lahars, there are some advantages of living near a volcano. One of these is that volcanic ash, when broken down and washed off the flanks of a volcano, produces a particularly rich farming soil in the surrounding plains, which is simply replenished by the occasional shower of ash. It is hardly surprising, then, that some farming communities thrive on volcanic islands. When Soufrière in St Vincent erupted in 1812 there was a fear of famine, because the fields had been destroyed. But crops planted in the freshly fallen ash grew so quickly that disaster was averted. Subsistence farmers in Papua New Guinea even perform 'rain dances' to bring about ash falls.

Secondly, some valuable chemicals are concentrated and deposited by volcanoes.

Volcanic Advantages *Volcanoes produce fertile soils (top), deposits of minerals such as sulphur (middle), and heat for geothermal power stations (bottom).*

When water is heated to temperatures way above its normal boiling point by magma in the pressure-cooker conditions deep inside the Earth's crust, it is highly corrosive, and can dissolve minerals from the rocks through which it passes. These minerals are then deposited at the Earth's surface as the pressure is released and the water cools. In particular, deposits of sulphur and mercury are exploited industrially in volcanic areas.

Thirdly, there is the potential for geothermal power. Water heated at depth within a volcano can be brought to the surface and used to drive turbines, producing electricity. Italy, for example, has been exploiting its own geothermal heat since 1904, and California and New Zealand have had geothermal power stations for decades. Iceland even uses hot volcanic water directly to heat buildings; take a shower in a hotel in Reykjavik and the smell is distinctly sulphurous.

Hawaii is now establishing its own geothermal energy programme, despite a setback in 1991 when exploratory drilling hit a pocket of steam that was far nearer to the surface than anticipated. The engineers were unprepared and the result was a blowout, with steam screeching upwards from the damaged well, and contaminating the surroundings with foul-smelling volcanic gases and overpowering noise. This was a timely warning that even normally placid basaltic volcanoes can be unpredictable.

Natural Pollution

It is not necessary to live on a volcano's flanks to be affected by its activity. The great quantity of gas and dusty solids ejected from an erupting volcano can have a significant effect on the Earth's atmosphere, creating a change of climate or weather conditions. The results of such activity have been widely studied, particularly in the mid 1980s when it was feared that new and more powerful nuclear weapons, if used, could have precisely the same effect – a 'nuclear winter'.

Some of the environmental damage that is commonly thought to be man-made can also be caused by volcanic activity. Acid rain, for example, is a serious pollution problem, especially in northern Europe and North America. Industrial gases, particularly sulphur dioxide, emitted from factory chimneys can combine with moisture in the air to form dilute acids, such as sulphuric acid, which then fall as rain, killing off freshwater life in lakes and damaging trees and buildings. But sulphur dioxide is one of the most abundant volcanic gases, too, and acid rain is a frequent side effect of eruptions. Poás volcano in Costa Rica has a crater that is lined with yellow from the sulphur of its eruption, and fumaroles – vents that emit volcanic gases – steam away

Poison Gas *The gases given off by volcanoes, such as these from fumaroles in New Zealand, can be toxic.*

around it. From time to time, the volcano erupts – not particularly violently, but blasting out sulphur-rich clouds of gas that fall as acid rain and destroy the crop of the coffee plantations that reach up the flanks of the mountain.

The Earth's stratosphere contains a layer rich in a gas called ozone, which protects life on Earth from the harmful effects of the Sun's ultraviolet radiation. In the last decade or so, it has been found that atmospheric pollution by various gases is breaking down this ozone layer, and that the resulting health hazards such as skin cancer are increasing. However, it is not just human pollution that is to blame. In 1991 scientists found that the reactions which led to the breakdown of the ozone took place not in the atmosphere itself, but on the surface of tiny ice and dust fragments floating at these stratospheric altitudes. This discovery followed a decade of spectacular volcanic eruptions, including those of Mount St Helens in 1980, El Chichon in Mexico in 1982, and Unzen in Japan in 1991, in which a great deal of dust entered the atmosphere. The researchers found that the vast quantity of particles erupted by El Chichon would have been sufficient to destroy up to 7 per cent of the ozone in the ozone layer. The ozone layer continually renews itself, but even a temporary loss of this magnitude is cause for concern.

Airborne volcanic dust can constitute an immediate physical hazard too. Each particle is hard, angular and glassy. Large quantities can cause severe respiratory problems in

Deadly Rain *Gases erupted from Poás in Costa Rica give rise to acid rain that periodically destroys the coffee crop on the mountain flanks.*

Breathing Difficulties *When Mount St Helens erupted, airborne silica dust particles, like powdered glass, caused respiratory diseases in Washington state.*

AN APPROPRIATE MEMENTO

The eruption of Mount St Helens in the early 1980s spread ash over much of Washington state, killing thousands of wild animals and sometimes causing severe disruption to traffic and everyday life. One enterprising artist found that the glassy ash could be melted down and moulded into souvenirs – ashtrays, as it happened.

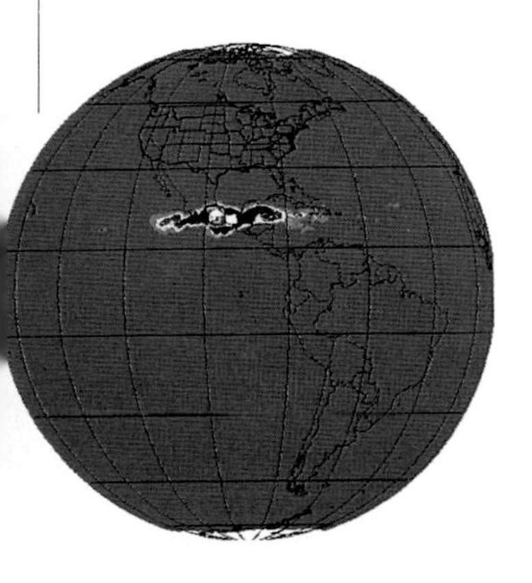

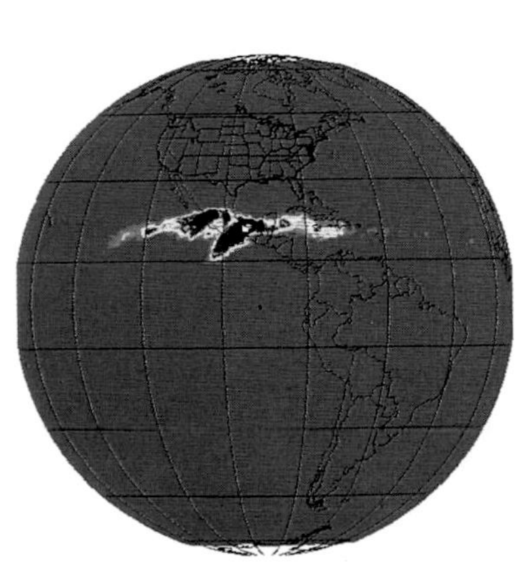

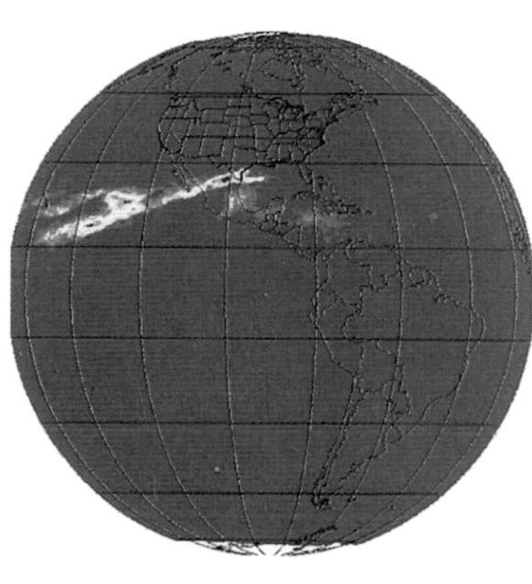

Seen from Space *The gas and dust plume from El Chichon was plotted by satellite, as it spread across the Caribbean and the Pacific during the first five days of eruption.*

Engine Failure *Dust from the eruption of an Alaskan volcano in 1989 clogged the engines of a 747 jet and almost brought it down.*

human beings, and mechanical problems in machines. In December 1989, a KLM 747 jet with 245 passengers was making its descent towards Anchorage, Alaska. Redoubt volcano had been erupting for several days and an ash plume had risen some 40 000 ft (12 km) into the air. Despite a warning by the Federal Aviation Administration, the KLM flight continued its approach; as it entered the cloud, all four engines failed. The aircraft, which suffered serious damage because of the abrasive effect of the dust, fell 13 000 ft (4 km) in eight minutes before the crew managed to restart the engines and make a safe landing at the airport.

Climate Changes

When scientists analysed ice cores drilled from the Greenland ice cap, they discovered that layers of ice laid down about 210 BC were extremely rich in sulphur. From this they deduced that an eruption had taken place in Iceland at around that time. The date also marks the onset of abnormally cold winters in Europe and a famine that wiped out half of the population of northern China. It seems likely that the dust and gas blasted into the atmosphere by the eruption blocked much of the sunlight that would otherwise have reached the Earth's surface, and produced cooling in the Northern Hemisphere.

The eruption of Tambora, on the island of Sumbawa, in Indonesia in 1815 is regarded by historians as one of the greatest in recorded history. On April 10 and 11, the island was shaken by explosions that were heard in Sumatra, 900 miles (1500 km) away. Between 7 and 36 cu miles (30-150 km^3) of debris were blasted into the sky, but the precise details are vague, since no one survived to tell the tale. Falling ash spread 300 miles (500 km) to the west and 190 miles (300 km) to the north, choking the seaways with a layer of floating ash known as pumice. Crops on the surrounding islands were obliterated and famine, along with disease, accounted for the vast majority of the 92 000 death toll. The aftereffects were more widespread, as ash circulated in the Earth's stratosphere. There was a worldwide drop in temperature, and the following year was known throughout Europe as the 'year without a summer'.

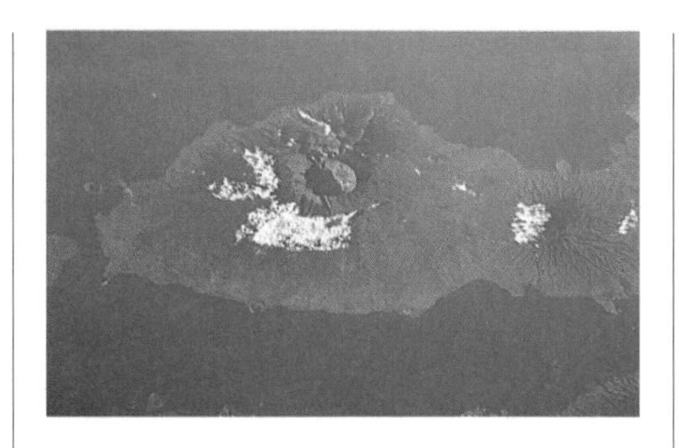

After the Eruption *From space, the peaceful crater of Tambora now shows little evidence of having generated in 1815 the biggest explosion in recorded history.*

COOLING THEN WARMING

In some instances, paradoxically, the cooling effect of volcanic dust in the atmosphere may lead to warmer winters. A temperature drop in the atmosphere over the ocean can increase the strength of the winds blowing from the oceans over the land. The moist air they bring with them moderates the climates of coastal areas – cooling them in summer and warming them in winter. If the winds increase, so does the climate modification. This might explain the mild weather in Europe and North America in the winter of 1991-2, after the eruption of Mount Pinatubo in the Philippines.

Volcano Country *The Virunga region of Zaire (right), with its mountains, glaciers and volcanoes, is so spectacular that it was declared a national park in 1925.*

Red Sky at Night *J.M.W. Turner's angry skies may reflect the effect of dust particles that were discharged into the atmosphere by the Tambora eruption of 1815.*

THE SHUDDERING EARTH

When forces deep within the Earth push the rigid crust beyond endurance, the rocks can no longer take the strain, rupturing and shearing in a massive burst of energy. The result is an earthquake that can lay whole cities flat.

For a decade, the inhabitants of China thought that the recurring problem of earthquakes in the People's Republic had been solved. After the devastating Xingtai earthquake of 1966, the Chinese Government established a comprehensive network of earthquake monitoring stations throughout the country. Before long, they had learned to recognise many kinds of advance warnings and were confident that they could predict earthquakes. The system came on stream just before the great Haicheng earthquake of 1975, and it was a resounding triumph: the earthquake was predicted and the population evacuated. Many lives were spared and the international community hailed the exercise as a thorough success.

But in July 1976, when the people of Tangshan noticed peculiar lights in the sky, goldfish jumping out of tanks, and fluctuating water levels in the wells, no earthquake warning was issued by the newly proved and infallible system. At 3.42 am local time on July 28, those who were up and about heard an ominous rumbling as a geological fault, 7 miles (11 km) directly below Tangshan city, began to shear. The ground surface shuddered for a few seconds, then suddenly moved in a clockwise twisting motion. This was followed immediately by an anticlockwise twisting motion, accompanied by vertical jolting. By the third twisting movement, the buildings were beginning to collapse, and

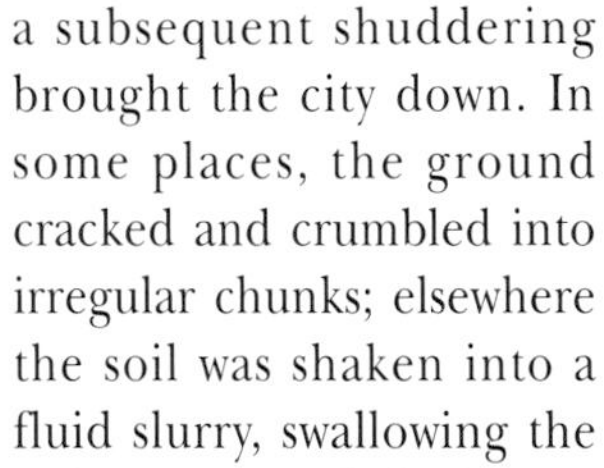

a subsequent shuddering brought the city down. In some places, the ground cracked and crumbled into irregular chunks; elsewhere the soil was shaken into a fluid slurry, swallowing the foundations of otherwise sturdy buildings. Ten seconds after it had started, the shock died away. In that time, almost the entire city became a level waste of rubble and 242 000 people were dead or dying. The earthquake had a magnitude of 7.8 on the Richter scale; the following afternoon, an aftershock, of magnitude 7.1, struck the same area, destroying much of what had been spared by the main shock. It was the worst earthquake disaster of modern times.

Only one house in 50 survived, and only one factory in ten was left standing. Seven moving trains were toppled from their tracks. All four city hospitals were flattened. Many people – about 16 per cent of the victims – were hit by falling masonry, but 30 to 40 per cent died of suffocation from the dust of collapsed brick and adobe buildings, as they lay trapped in the rubble of their own homes. Others – 20 to 30

SOIL SPLIT *The ground and buildings were cracked by the Haicheng earthquake of 1975, but most of the population was evacuated in time.*

EARTHQUAKE ZONE *In earthquake-prone areas, such as Los Angeles, architects favour columns that buckle rather than collapse (right), and floors and walls that do not crumble into a deadly rubble.*

EXIT-ONLY
WRONG WAY
STOP
SEVERE TIRE
DAMAGE

Industrial Damage *Little remained of the locomotive factory after the most disastrous earthquake in recorded history had flattened the Chinese city of Tangshan, killing 242 000 people.*

per cent – died later through exhaustion, hunger, thirst or lack of medical care due to inevitable delays in their rescue. Communications were badly hit. Some 250 miles (400 km) of railway track were put out of action, through subsidence or bending, as bridges collapsed and trains toppled. All roads were destroyed by cracking or by soil slurry. The telephone and radio systems were disrupted, and it took several hours for the first news of the disaster to reach the Chinese capital, Beijing.

The Rescue Operation

Relief work was enthusiastic but disorganised, as people immediately started digging with their bare hands in the rubble for trapped victims; some say that 80 per cent of those who were trapped, but survived, had been rescued by their neighbours immediately. Rescue workers were brought into the area but with little in the way of equipment, or even of food and drink for themselves. Medical attention for the survivors was meagre.

The liquefied soil clogged the wells, and many water pipes, drains and sewers were destroyed. Since the weather at the time was particularly hot, disease was inevitable. Enteritis and dysentery reached their peak about a week after the earthquake. Influenza,

typhoid and encephalitis then set in. Compounding this was the haphazard way in which bodies were buried – usually in shallow graves close to their original homes. One of the first tasks of the official rescue teams was to remove these bodies to fresh burial sites, away from areas where they could cause infection, and to sterilise the original graves. Emergency vaccination campaigns prevented the further spread of diseases – indeed, the mass vaccination was so thorough that the incidence of many diseases the following year was lower than it had been before the earthquake.

Rebuilding Tangshan

Tangshan was a big mining and steel centre, and once survivors had been rescued, the Government's priority was to re-establish the industry. Strangely, the underground shafts and tunnels of the coal mines suffered little damage. The machinery, however, was badly hit and the mines were severely flooded. Some 30 000 miners on the night shift during the night of July 28 had survived, but it was two days before they reached the surface and saw the appalling state of their city. By early 1977, seven of the eight mines in the area were back in action. Workers from all over the country were drafted in to repair the damaged steelworks. By May 1977, steel was being produced again at pre-earthquake levels, and the economy of the area was well on the way to recovery.

In September 1976, shortly after the earthquake, the decision was made to rebuild the entire city on the same site. Until the earthquake, Tangshan had been one of the most polluted cities in China, having grown piecemeal as the industries developed. The rebuilding was carried out with the benefit of modern urban planning. Eight years later, after an expenditure of 24 million yuan ($7.7 million), the city was once again a thriving industrial centre.

RISING FROM THE RUINS
It took eight years to restore Tangshan to a prosperous industrial city after the devastation of the 1976 earthquake.

Energy Release

Earthquakes are caused by the release of tension in the rocks of the Earth's crust and upper mantle. As a rule, this tension is released in one very brief burst of energy after it has been building up for a long period of time – decades or even centuries.

The point of origin of an earthquake – the area in which that tension is released – is called the focus. It is here that the greatest disturbance occurs, although 'point' is something of a misnomer, since the focus may actually represent a volume of several cubic miles. The point on the Earth's surface directly above the focus is called the epicentre, and it is here that most of the damage is done.

Most earthquakes occur at active faults, or cracks, in the Earth's crust, where movement has taken place. These faults develop as moving portions of the crust grind up against each other. The movement is not steady – the rocks on either side of the fault are locked together by the sheer solidity and mass of the Earth's crust – but eventually the tension becomes too great. The two sides of the fault snap into their new positions and settle. This movement, involving immense quantities of rock, radiates shock waves outwards like the ripples in a pool. Usually the movement is so great that the rocks on either side of the fault overshoot, and so the resulting condition is still

AVERTING A GREATER DISASTER
China's plains are so prone to flooding that, after every earthquake, soldiers and civilians are called upon to reinforce the river banks.

unstable. Over the next few days or weeks, the rocks will shuffle back and forth until they find a new equilibrium. It is these later movements that produce the aftershocks associated with earthquakes.

Shock Waves

Blasting out from the earthquake's point of origin and spreading through the crust of the Earth, these shock waves are the direct cause of an earthquake. One set of waves, known by geologists as P-waves or Pressure waves, push and pull the particles of rock back and forth in the direction in which the wave travels. Another set, known as S-waves or Shear waves, cause the particles of rock to vibrate up and down.

Since P-waves travel through the Earth's crust approximately twice as fast as S-waves, they arrive earlier at recording stations (although they would have set out at the same time). It is this difference in arrival times that enables geologists to estimate their distance from the point of origin.

Earthquake Rift *Ground is disturbed by an earthquake.*

CAUSES OF AN EARTHQUAKE

Earthquake Anatomy *The focus of a destructive earthquake is usually quite shallow – less than 43 miles (70 km) below the surface – with the greatest damage tending to occur at the epicentre.*

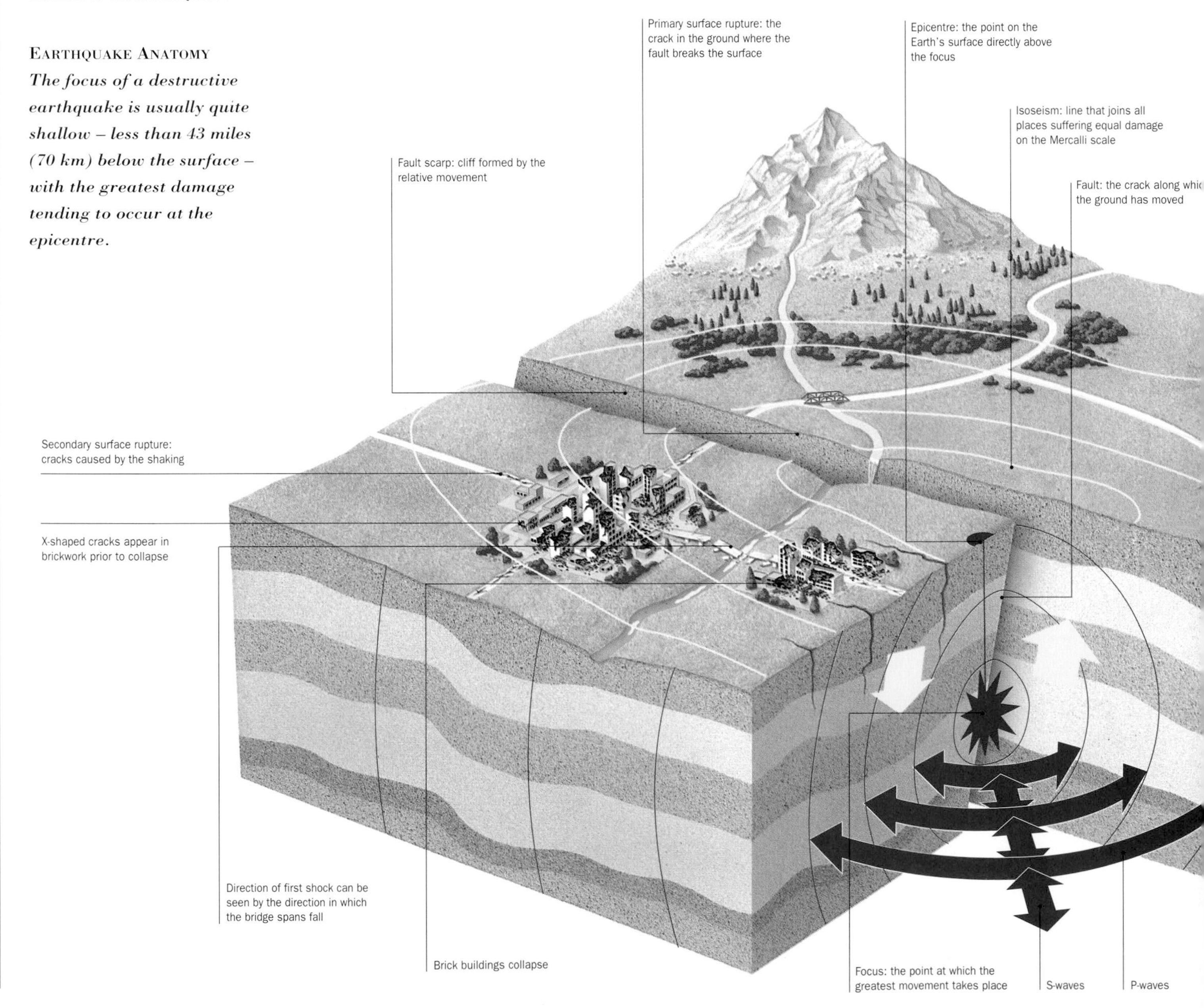

CORNER CHAOS *Buildings collapse in San Francisco.*

HIGHWAY WRECK *A multilevel expressway collapses.*

CLEARING UP *Buildings of medium height tend to survive.*

One seismographic station can tell that an earthquake took place at a particular distance away, but has no means of ascertaining where. A second seismographic station will likewise produce an estimate of its distance from the earthquake, and the two sets of readings can then be used to pinpoint the origin of the waves. The readings from a third seismographic station can then be used as further confirmation.

P-waves and S-waves pass swiftly through the Earth, but are detectable only by instruments. There is, however, a third set of earthquake waves, known as L-waves, or Long waves, which travel more slowly over the surface of the Earth, and are responsible for the damage that we generally associate with an earthquake.

Tangshan, at the epicentre of its earthquake, experienced the full destructive force of the L-waves. While the P-waves and S-waves were sending their messages to the world's seismographic stations, the city was seized by the chaotic twisting and heaving of the Earth that was the result of the L-waves' surface motions.

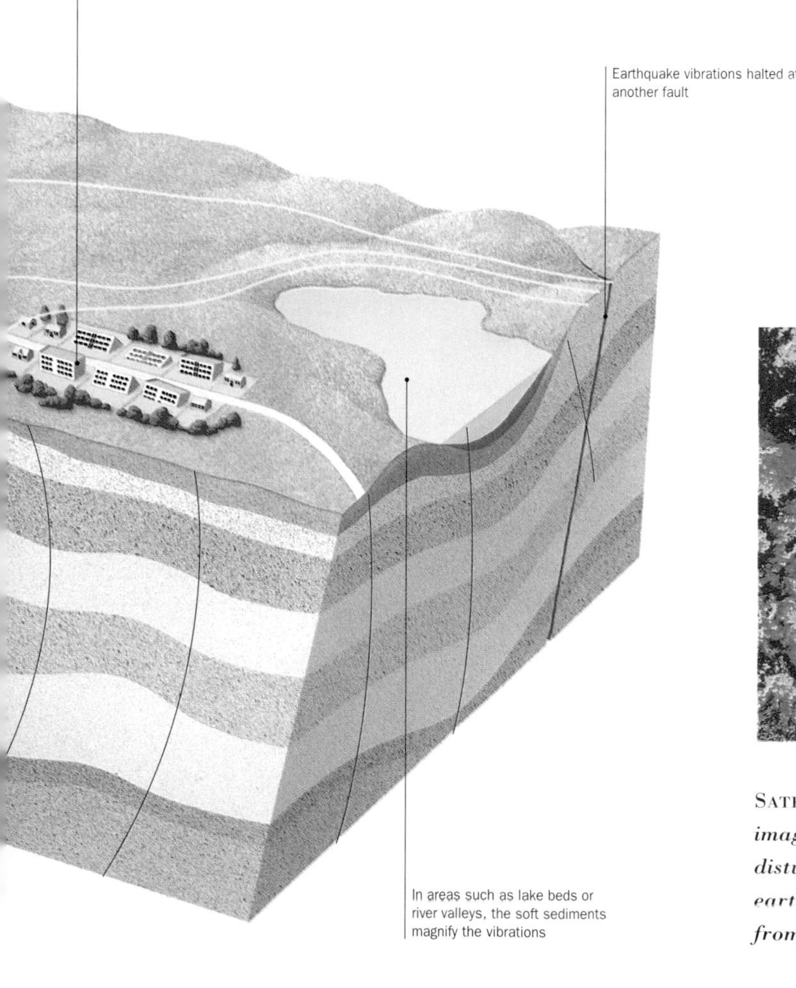

WHERE EARTHQUAKES HAPPEN

The surface of the Earth consists of individual plates that are constantly on the move, and it is at the edges of these plates that volcanoes are likely to be found. Much the same is true for earthquakes. The world's two major earthquake belts correspond to destructive plate margins, where the edge of the plate is being worn away. One of these extends around the Pacific Ocean, from New Zealand, up through the Philippines, Japan, Kamchatka, the Aleutians, Alaska, and then down through California, Central America and the Andes – the so-called 'ring of fire'. The other stretches westwards from the edge of the ring through the East Indies, up through Malaysia, and around the rim of the Himalayas, before going on to link up the Caspian, the Black and the Mediterranean Seas like dots in a puzzle.

SATELLITE VIEW *False-colour imaging shows the pattern of disturbance after an earthquake, as waves spread from the active fault.*

Constructive plate margins,

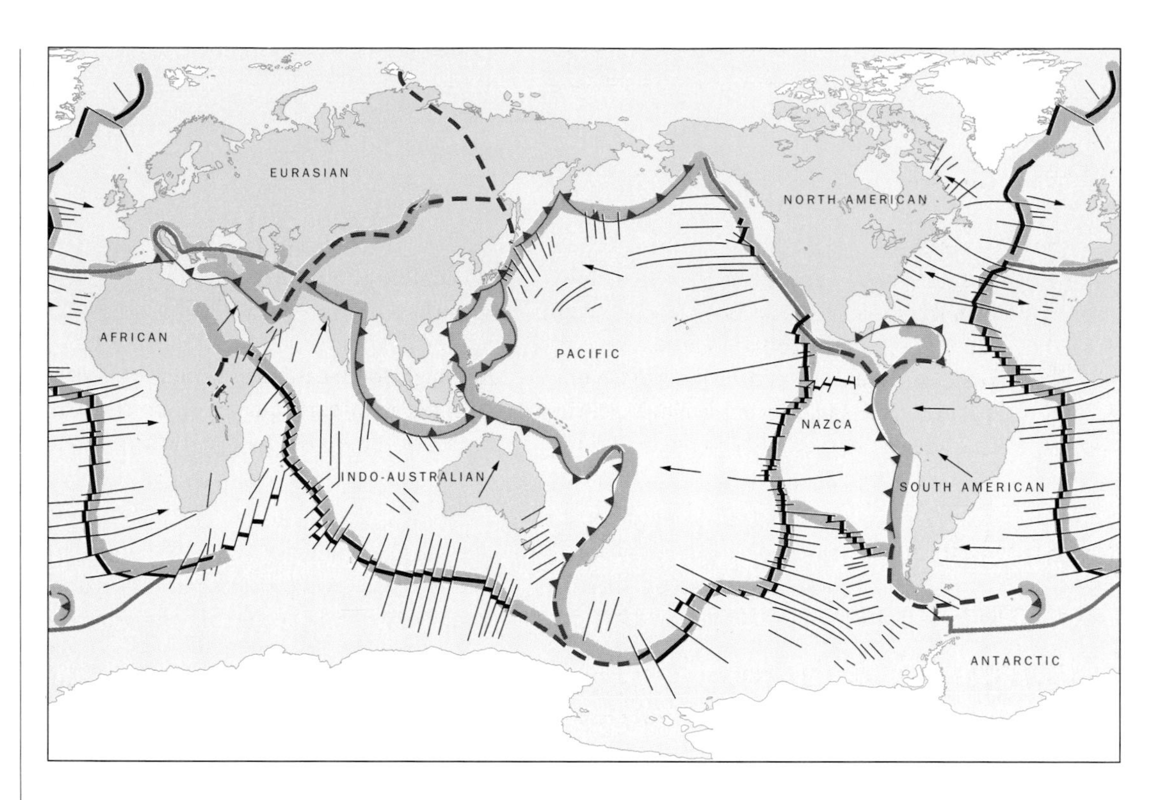

TECTONIC FEATURES

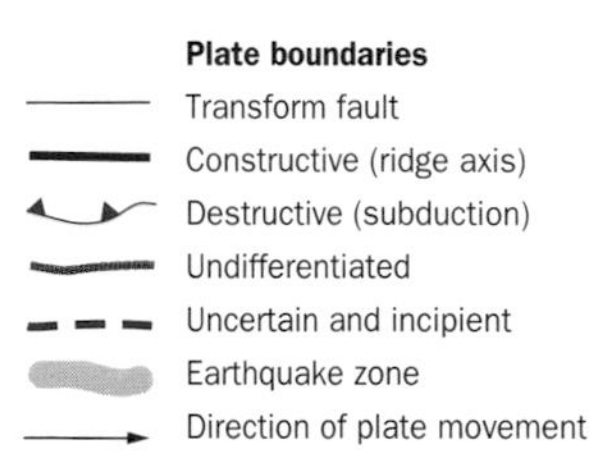

Earthquake Zones *Like the distribution of volcanoes, the location of earthquakes follows the plate boundaries on the Earth's surface. It is the movement of the plates along these boundaries that produces earthquakes.*

on the other hand, usually lie along the bottom of the oceans, and earthquakes here tend to be overlooked. Exceptions, however, include the Great Rift Valley, which stretches from the valley of the River Jordan southwards through East Africa, and Lake Baikal in eastern Asia. Both of these lie on constructive plate margins within a continental landmass, and both have their own minor earthquake belts.

Different Depths

Earthquakes can be categorised by their depth. An earthquake is regarded as 'shallow focus' if its focus lies less than 43 miles (70 km) below the Earth's surface; 'intermediate' if the focus lies between 43 and 186 miles (70 and 300 km) below the surface; and 'deep focus' if more than 186 miles (300 km). The deepest focus yet recorded, below the East Indies in 1934, was 450 miles (720 km).

At a typical destructive plate margin, the shallow focus earthquakes generally occur close to the ocean trench itself, where the plate is shallowest, while deeper earthquakes tend to be farther away from the trench, on the inside of the volcanic island arc, where the plate is deepest. This distribution is consistent with the fact that the ocean plate is curving down and descending into the Earth's mantle at an angle of 45 degrees. Usually, it is the shallow focus earthquakes, like the Tangshan earthquake, that wreak the worst havoc because deeper ones may be muffled by the sheer mass of rocks surrounding them. The deepest to cause any significant loss of life (1000 people were killed) took place in Romania in 1940 – the focus at a depth of around 100 miles (160 km).

In many places, such as California, the plates are sliding not beneath, but past one another. Along the notorious San Andreas Fault, the coastal region of California is shearing away northwards at such a rate that in 11 million years' time the area of Los Angeles will have drawn level with San Francisco. The San Andreas Fault – or, rather, system of faults – produces constant shallow focus earthquakes, of which the most famous perhaps is the one that flattened San Francisco in 1906, killing 700 people.

Images of Disaster *The 1906 San Francisco earthquake was one of the first to be extensively recorded for posterity by photography.*

SCARRED EARTH *The San Andreas Fault extends almost the full length of California and is responsible for major earthquakes.*

But why did the most deadly earthquake in history take place in Tangshan, hundreds of miles from any plate margin? It is possible that such earthquakes, known as 'intra-plate earthquakes', occur along old faults, left over from some ancient tectonic activity. Earthquakes of this kind can be surprisingly strong and sudden, perhaps because they are not associated with any well-defined fault system that has been releasing tension over the millennia.

INTRA-PLATE EARTHQUAKES

On December 28, 1989, an 11-second earthquake measuring 5.5 on the Richter scale surprised the inhabitants of Newcastle near Sydney, New South Wales, killing 12 people. Although south-eastern Australia has been afflicted with earthquakes since records began in the middle of the 19th century,

THE MEDITERRANEAN EARTHQUAKE ZONE

The northern edge of the African plate has been grinding away at the southern edge of the European plate for about 50 million years. Before that time, a great embayment, or landlocked area of water, called the Tethys Sea, lay between them. When the two plates closed in on one another and collided, the Tethys Sea was squeezed out; and now all that is left of it are the Mediterranean and a string of drying puddles: the Black Sea, the Caspian and the Aral Seas.

Even now, the African plate is moving eastwards in relation to the European plate and, as it goes, the mountain ranges in between are being twisted into great S-shapes – from the line of the Atlas Mountains of North Africa across the Mediterranean via Malta and Sicily, up through the Apennines of Italy, around the curve of the Alps and down the Dinarics, swinging up again around the Carpathians. The great square mass of the Iberian Peninsula used to be tucked into the Bay of Biscay, but it was plucked out and swung around, with the Pyrenees crumpled up at the hinge. The islands of Corsica and Sardinia were once part of the European landmass, snuggling into the bays where Marseilles and Genoa now stand, but they too have been wrenched out and dragged into the ocean.

With all this activity, it is hardly surprising that the Mediterranean region is a highly active tectonic zone. Famous volcanoes abound – Vesuvius, Etna, Stromboli, Vulcano (which gives its name to the phenomenon). Earthquakes rock the region, and early recorded history mentions them frequently. Roman basreliefs show tumbling buildings; legend tells of the destruction of Atlantis, which some scholars identify with the Minoan civilisation of Crete; and the Bible describes 'a trembling in the host, and in the field'.

The time may come when there will be no longer be a Mediterranean Sea. The whole region may be crushed up into a mountain range comparable to the Himalayas as they are now. But that will not be for at least another 50 million years.

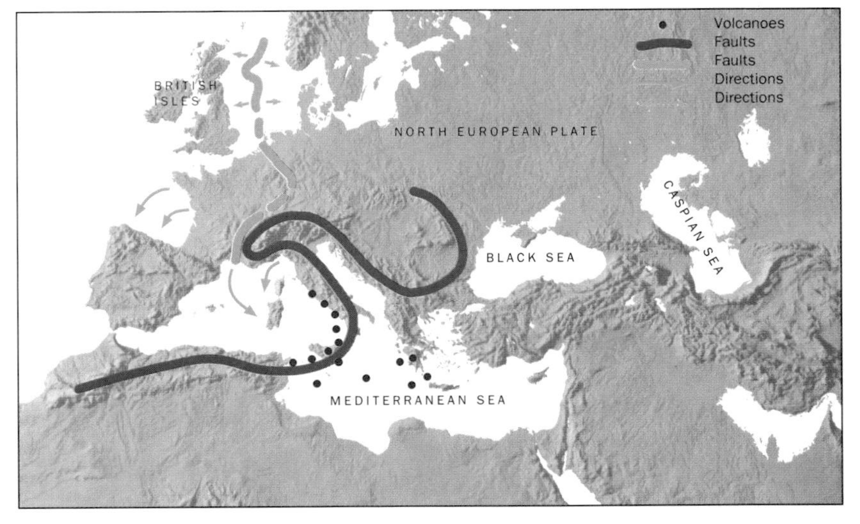

PLATE TECTONICS IN ACTION *Europe twists and crumples under the onslaught of Africa.*

SICILIAN SKY *Mount Etna is near the boundary between the European and African plates.*

this was the first fatal incidence in the area, which is far from any plate boundary. Likewise, there appears to be an earthquake zone in Missouri, along what seismologists call the New Madrid Fault Zone. This is right in the middle of the continent, where scientists would least expect earthquakes, yet the New Madrid area experiences about 200 minor earthquakes every year. Old continental interiors have always been regarded as solid and stable. Nowadays, geophysical investigation shows that they are riddled with cracks and old fault systems, and that there is still the potential for a great deal of tectonic activity a long way from plate margins.

MAN-MADE EARTHQUAKES

In 1935 the Boulder Dam (now called the Hoover Dam) across the Colorado River was opened, to be hailed as a great human achievement – part of a grand plan to bring civilisation to the wilder corners of the American Far West. The following year, however, a series of earthquakes began. Although these were not big, no earthquake had been recorded in the area before. The earthquakes, several hundred of them, continued until 1940, by which time the lake behind the dam – Lake Mead – had filled to its working capacity: at which point the earthquakes stopped. It was as if the filling of the reservoir had put strains on the surrounding crust – strains that manifested themselves as earthquakes – after which, things settled.

This was the first recorded instance of human engineering being implicated in the generation of earthquakes. Since then, similar phenomena have been observed at other artificial lakes, particularly in Greece and France. Until the Koyna Reservoir disaster in 1967, however, these were regarded as an academic curiosity.

THE ROMAN VIEW

Nearly everybody in ancient times who experienced earthquakes believed that they were the work of angry deities. It was the Roman writer Seneca (4 BC-AD 35) who gave the first recorded opinion that they might have been purely physical phenomena, arguing that they were caused by the movement of vapours in the Earth's interior.

DANGER DAM *The water behind the Hoover Dam seeped into cracks, lubricating old faults and producing a series of earthquakes in the 1930s.*

The Koyna Reservoir, built on stable old continental rocks in the heart of the Indian continent, was filled between 1962 and 1965. Almost immediately, earthquakes began to affect the area; they increased in frequency and strength, culminating in a disastrous earthquake that measured 6.4 on the Richter scale, killed 177 people and injured 2000 more. What happened at Killari, in the central Indian state of Maharashtra in 1993, was much worse. Earthquake hazard maps divide an area into regions, according to the likelihood of earthquakes. The hazard map of India shows nothing alarming, except in the region along the edge of the Himalayas – a destructive plate margin – where the greatest earthquake in the sub-continent's history occurred in 1934. The central region in which Killari is situated shows up as the least likely to experience an earthquake. At Killari something quite new for the area occurred. At 3.45 am on September 30, 1993, an earthquake of magnitude 6.4 struck this rural but densely populated area, flattening 20 villages and killing some 10 000 people. The buildings were made of stone and mud – the most vulnerable architecture in the event of an earthquake. The Lower Tima Reservoir, 10 miles (16 km) upstream, was being filled at the time, and it is thought that this triggered the earthquake.

By the early 1980s, it had become apparent that at least a hundred reservoirs had

RESERVOIR TO BLAME? *The Indian town of Killari was flattened by an earthquake which may have been induced by the waters of the nearby Lower Tima Reservoir.*

given rise to earthquake activity. The phenomenon is still not fully understood, but the most plausible explanation is that water seeps along faults that had hitherto been dry, and lubricates them, rather like oil seeping around a rusty bolt. Movement becomes easier and tensions that have been building up for millennia are more likely to be released.

Underground nuclear explosions can also generate earthquakes. A nuclear blast is not, in itself, an earthquake, and produces a completely different kind of seismographic reading. Shock waves spread outwards as an expanding sphere, whereas in an earthquake the shock waves will initially be generated in a particular direction. After a nuclear test explosion, any faulting in the area is shaken loose, and an underground blast is usually followed by aftershocks that are true earthquakes.

A curious piece of research conducted in 1994 found that the 13 major earthquakes that hit Japan between AD 684 and 1946 all occurred at times when the atmospheric pressure was high. At such times, pressure would be exerted on the exposed continental plate, but not on the oceanic plate that was sliding down beneath it. The difference in pressures might just have been enough to trigger an earthquake that was already building up in the area. This matches another piece of research which indicates that the lowering of the sea level during the Ice Age, when vast quantities of seawater were frozen in the ice caps, led to an increase in volcanic activity, as the water pressure above undersea volcanoes diminished. Conversely, volcanoes that rose above sea level, such as those in Iceland, would have been capped by thick layers of ice during an ice age. When the Ice Age ice caps melted and the pressure on the underlying volcanoes was removed, there were major phases of volcanic eruption – as the lava flows, dating from between 8000 and 4500 years ago, around the calderas of Askja and Veidirotn suggest. Another piece of research has revealed that, in the last 30 years, the onset of an irregular change in the circulation patterns of the Pacific Ocean – a phenomenon known as El Niño – coincides with increased earthquake activity along the East Pacific Rise, the ocean ridge that marks the constructive plate margin of the Pacific, in the area of Easter Island.

CENTURIES OF SHAKES
The island chain of Japan was built by plate movements, and these same forces continually rack the country, as this 1891 photograph of an earthquake in Wakamori shows.

MEASURING EARTHQUAKES

There are two scales on which an earthquake can be measured. The best known is the Richter scale, which measures the earthquake's magnitude; the other is the Mercalli scale, which measures its intensity.

The magnitude of an earthquake is measured by calculating and comparing the sum of energy released on a scale of 0-10, compiled by the American geophysicist C.F. Richter in 1935 (the highest any earthquake has registered is 8.6). The measurements are based on readings from a seismograph located 62 miles (100 km) from the epicentre.

Since the Richter scale is logarithmic, a magnitude 4 earthquake, for example, releases only about one-thirtieth, and not four-fifths, of the energy released by a magnitude 5 earthquake. Beware, then, if you hear a news report that begins: 'An

Sinking Buildings *When soil is turned to fluid by an earthquake's shock waves, buildings may sink into the ground, as in this earthquake in Japan in 1964.*

aftershock, almost as great as the original earthquake, has struck . . .' News reporters have almost certainly been comparing Richter scale measurements and misinterpreting them – for aftershocks are usually far smaller than the initial earthquake.

Perhaps a more useful concept for those who actually witness an earthquake is provided by the Mercalli scale, a more subjective measurement, based on observed phenomena and levels of damage. Measurements differ from one part of a stricken area to another, with the highest reading generally near the epicentre. A particular earthquake will give a single reading on the Richter scale, but a whole range of them on the Mercalli scale. For example, the earthquake of Boxing Day, 1979, in Longtown in the Scottish Border country (an intra-plate earthquake – the Scottish-English border is a long way from a plate boundary) measured 4.8 on the Richter scale. On the Mercalli scale, it measured IV in Annan, 12 miles (19 km) from the epicentre (in fact, it woke the author of this book, and the disturbance of sleep is one of the definitions of point IV). In Carlisle, 7 miles (11 km) from the epicentre, it qualified for a measurement of VI by bringing down chimneys. The scale in use today is actually a modified Mercalli scale.

Building for Earthquakes

Traditional Japanese dwellings are ideally suited for an earthquake zone. The bamboo from which they are built simply twists and turns with the shaking earth and the paper panels fall out without causing physical injury. Afterwards, the building can be reassembled relatively straightforwardly. However, such building materials do not necessarily suit 20th-century requirements. City buildings, office blocks and factories are built on a scale beyond the capacities of bamboo and paper, although their design in earthquake-prone areas, such as Japan and California, must take the effects of earthquakes into account.

Engineers study the effects of earthquakes on structures in the laboratory. For example, the National Center for Earthquake Engineering Research at New York State University, Buffalo, has designed a 'shake table' which simulates the effect of an earthquake and is used to test various construction methods by means of scale models. One of the first lessons from such experiments was that tall buildings tend to whip back and forth during an earthquake. If the dimensions of the building are such that the speed of the whipping action happens to be 'tuned' to the vibration of an earthquake (the way a piano string may hum when you sing to it), the whipping intensifies and the building collapses.

This was exactly what happened when an earthquake of magnitude 8.1 struck Mexico City in 1985. Buildings higher than about five storeys, and lower than 15 storeys were badly damaged, because they seemed to have a natural resonance that coincided with and magnified the vibration of the earthquake. Those below five and above 15 storeys seemed to have come off more lightly, Japanese engineers have

THE MERCALLI SCALE

The original was compiled by the Italian seismologist Giuseppe Mercalli in 1902, but it has gone through a number of revisions since then. Currently, it runs as follows:

- I Only felt by instruments.
- II Felt by people at rest, especially on upper floors. Suspended objects may swing.
- III Felt indoors. Vibrations like passing traffic.
- IV Many people feel it indoors, a few outdoors. Crockery and windows rattle. Standing cars rock. Some sleepers awake.
- V Felt by nearly everyone. Tall objects rock. Plaster cracks.
- VI Everyone feels it, many are frightened. Chimneys fall. Furniture moves.
- VII Most people run outdoors. Damage to weakly constructed buildings. Felt by people in moving vehicles.
- VIII Considerable damage to most buildings. Heavy furniture overturned. Some sand fluidised.
- IX Even well-designed and sturdy buildings badly damaged, moved from their foundations. Ground cracks. Pipes break.
- X Most masonry destroyed. Landslides occur. Water slops from reservoirs and lakes. Railway lines bend.
- XI Few structures remain upright. Bridges fall. Extensive fissures in the ground. Underground pipes totally out of action.
- XII Total destruction. Ground thrown into waves. Objects flung into the air. You would be lucky to survive this one.

been trying to turn this effect to advantage by installing gigantic pendulums in tall buildings. The theory is that once sensors in a building detect and analyse an earthquake, the pendulum is set swinging at a frequency that will cancel out the swing of the building caused by the earthquake.

Another way of coping with earthquakes is to separate a building from its foundation – a technique known as 'base isolation' by the Californian engineers who have been developing it since the 1970s. The building is set on a stack of rubber cushions designed to absorb the vibrations of an earthquake

THE SEISMOGRAPH

Scientists measure the disturbance caused by an earthquake with a device known as a seismograph. This instrument was invented by the British seismologist John Milne in the 1870s and consists of a weight suspended in a laboratory. When an earthquake occurs, the laboratory shakes, while the suspended weight remains stationary. The relative movement between the laboratory and the weight is recorded, usually on a rotating drum. As a rule, more than one seismograph is used to measure vertical and horizontal vibrations, and to calculate the magnitude.

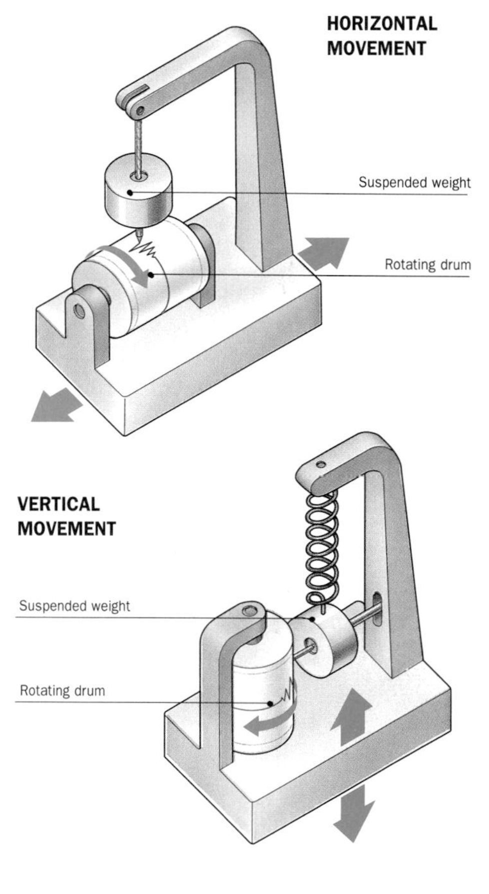

HANGING WEIGHTS *Vibrations in any plane are measured by the seismograph.*

SWINGING WITH THE VIBRATIONS *Buildings of different heights suffered different amounts of damage during the Mexico City earthquake of 1986. Those with the same natural resonance as the vibrations of the earthquake were the most vulnerable.*

TOKYO'S DISASTER DAY

After the devastating earthquake of 1923, Tokyo has been very serious about protecting its citizens against a recurrence. Every year, on September 1, the anniversary of the earthquake, the whole city undergoes intensive escape and rescue drills.

and to prevent them from reaching the building itself. In more advanced versions of base isolation, hydraulic pistons isolate the building.

Nevertheless, earthquakes will inevitably do damage to buildings, and there are more rudimentary ways of minimising this: concrete and brickwork can be encased in steel mesh, for example, so that if the structure does fail, the bits are held together and do not drop on the people below. All this costs money, and planners in earthquake-prone areas are obliged to weigh the likelihood of a major earthquake against the cost of precautions. In earthquake-prone areas, the first aim is safety – it does not matter if a building is unusable subsequently, so long as the residents survive. Beams and columns made from deformable materials – hollow steel tubes, for instance – bend and twist permanently under strain, but at least they absorb the worst of the shock as they do so, rather than break like more rigid structures.

Safety standards for buildings are revised after every major earthquake. In the first standards set in the 1920s, buildings had to withstand ground forces equivalent to 0.1 times the force of gravity – the figure had doubled in Japan by 1950. The quantity of steel reinforcement required has also been increased over the years. When an earthquake struck Kobe in 1995, it was the oldest buildings that suffered most: 36 per cent of those built before 1971 were too damaged to enter, whereas the figure was only 11 per cent of those built between 1971 and 1980, and 8 per cent of those built later.

Predicting an Earthquake

If engineering for earthquakes is proceeding apace, the same cannot be said of earthquake prediction. The quest for a foolproof method of forecasting earthquakes has been under way for centuries, but with few repeatable successes.

One of the first stages in earthquake prediction is to identify what are known as 'seismic gaps' – the absence of earthquakes

New Town *The collapse of this overpass during the 1995 earthquake in Kobe showed the defects of some modern building techniques.*

OLD TOWN *Traditional wooden buildings in Kobe were vulnerable to fire (above left), but were less lethal than modern concrete types when they collapsed (above).*

in places where stresses and strains are known to be building up. Since the 1970s, scientists have been aware of a seismic gap in the San Francisco area, where major earthquakes have been expected for a long time. The fear is that when a major earthquake does occur, it will release stresses that have been building up for so long that the results will be devastating. The Loma Prieta earthquake, in a suburb of San Francisco in 1989, had a magnitude of 7.1 and killed 62 people. However, scientists did not regard this as the major release of tension they had been anticipating, and San Francisco is still waiting for the 'big one'.

In the shorter term, scientists look for any signs of strain. The most obvious symptom is the appearance of millions of tiny cracks in the earth, which may indicate that the ground has swelled. Satellite surveys, which nowadays work to very fine tolerances, are being developed to monitor very small movements of earthquake-prone areas. Tilt-meters measure how much the surface of the Earth tilts – which it does when an increase in the volume of the rock mass causes the ground to swell. Gravitometers detect differences in gravitational force between different areas of rock: if the rock

swells, it becomes less dense and a change in gravitational force can be detected. Changing density can also affect the rock's other properties, such as magnetism and electrical resistivity, and these changes can be detected on magnetometers and resistivity gauges. Long steel bars cemented into rocks can be monitored for deformation – a process known as stress-metering. Another idea in development involves the use of two parallel strands of optical fibre, one cemented firmly to a given area of rock and the other isolated from it. Light waves are transmitted in phase through both strands and, provided they remain in phase on arrival at the ends of each strand, a bright light is produced. If, however, there are stresses and strains in the rock, these will be replicated in the strand that is fixed into the rock, and the light waves passing through that strand will no longer arrive at their destination in phase with those arriving at the end of the isolated strand. Once the transmission falls out of phase, there is no longer a bright light – an indication that the area is under serious stress.

In the build-up to an earthquake, rocks

MATERIAL DAMAGE *Members of one of the most prosperous societies in the world – that of California – are periodically ruined by the randomness of earthquake strikes. Much of the research into earthquake prediction is taking place in the United States.*

TRANSPORT DISRUPTION *The destruction of the Santa Monica freeway in 1994 hindered the rescue operation.*

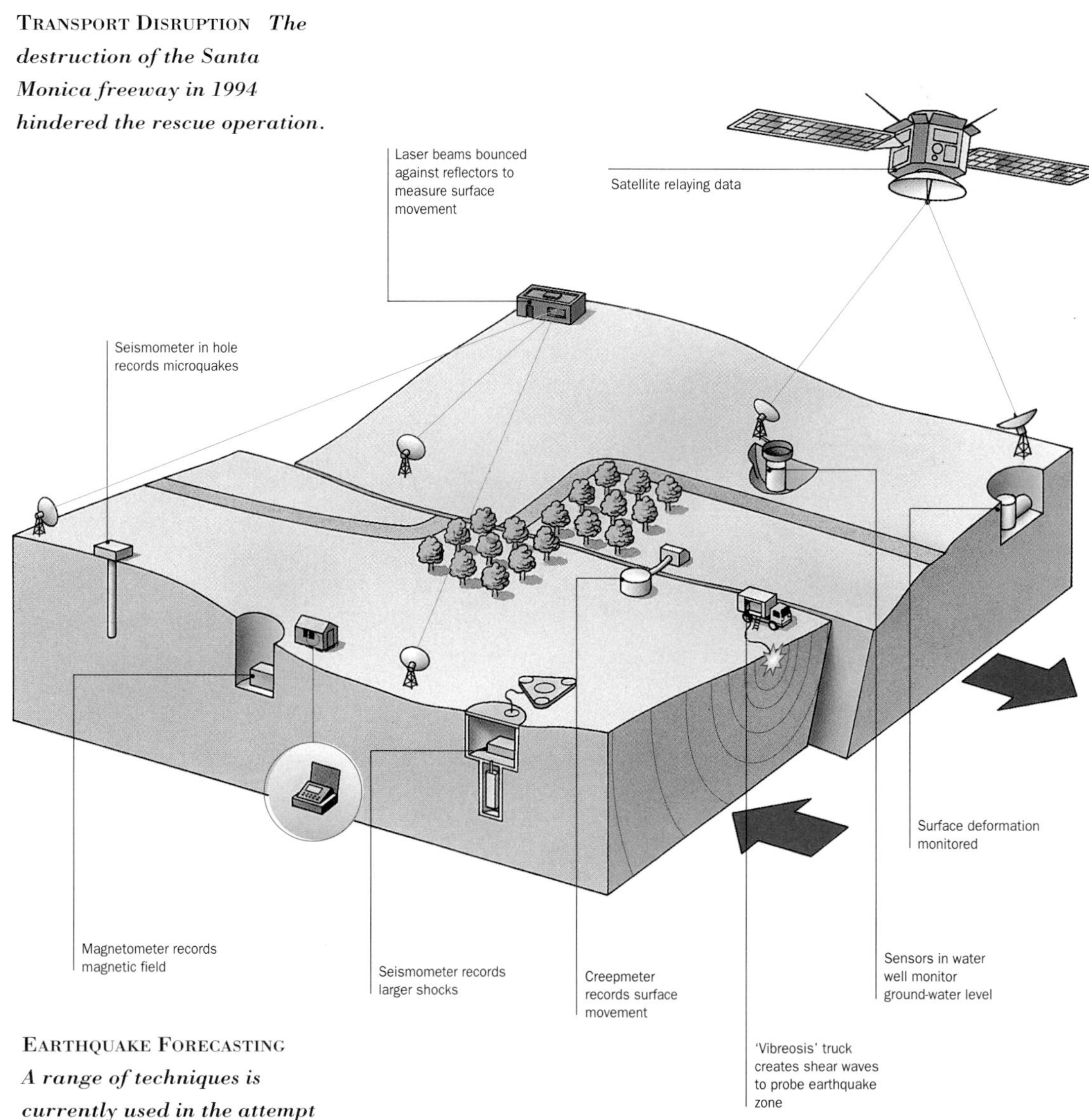

EARTHQUAKE FORECASTING *A range of techniques is currently used in the attempt to predict earthquakes.*

PREDICTIVE METHOD *At Parkfield in California, laser beams measure minute variations in the movement of the San Andreas Fault.*

crack and become more porous, facilitating the rise of fluids to the surface. These fluids give off gases, the most important of which is radon, because its radioactive nature makes it easy to detect. Others include hydrogen, helium and sulphur dioxide, and the levels of all these are monitored in earthquake areas by measuring their concentrations in ground water. Rising water can itself be an indicator of stress, and some earthquakes are preceded by changes in the level of wells.

Many less 'scientific' phenomena have been observed before an earthquake strikes. Sometimes, for example, a bright light is seen in the sky, which may be the result of a change in the electrical properties of the rocks. Electrical activity in trees has also been observed. There are also tales of strange animal behaviour. Snakes wake up from hibernation, pigs chew off their tails, and fish leap from the water. In China, bizarre events of this kind are taken seriously by the authorities, who combine them with more obviously scientific findings in their earthquake-prediction programme. This was the case with the Haicheng earthquake of 1975. At 2 pm on February 4, after a night of unusual phenomena and minor tremors, the people of southern Liaoning province were ordered from their homes into the open countryside. At 7.36 pm the earthquake struck. Two cities were flattened, but only 300 died, rather than the tens of thousands that might otherwise have been expected.

THE USE OF AN EARTHQUAKE

The shock waves caused by an earthquake can be used by geologists to determine the structure and composition of the Earth. P-waves, or push-and-pull waves, can pass through solids and liquids, whereas S-waves, or shear-waves, only pass through solids. Both types of waves are refracted, or bent, when they pass through materials of different densities. By looking at the pattern of waves received by seismographic stations in different parts of the world, scientists can tell the nature of the material through which they have passed. This is how we know, for example, that the Earth's inner core is solid, its outer core is liquid, and that the mantle and crust are solid, except for a zone near the top of the mantle that is rather softer: it is this soft layer that allows the plates to move.

MAJOR EARTHQUAKES

Date	Location	Death toll	Magnitude
1290	Gulf of Chihli, China	100 000	
1556	Shensi, China	830 000	8.3 (est.)
1667	Shemakha, Caucasia	80 000	
1692	Port Royal, Jamaica	2000	
1755	Lisbon, Portugal	60 000	8.6 (est.)
1797	Quito, Ecuador	40 000	
1835	Concepción, Chile	5000	8.5 (est.)
1835	Northern Japan	28 300	7.6 (est.)
1905	Jammu and Kashmir, India	19 000	8.6 (est.)
1906	San Francisco, USA	700	8.3 (est.)
1908	Messina, Italy	83 000	7.5 (est.)
1915	Avezzano, Italy	29 980	7.5 (est.)
1920	Gansu, China	100 000	8.6 (est.)
1923	Tokyo, Japan	200 000	8.3 (est.)
1927	Nan-Shan, China	200 000	8.3 (est.)
1932	Gansu, China	70 000	7.5 (est.)
1935	Quetta, India	30 000	7.5 (est.)
1939	Erzincan, Turkey	30 000	7.9
1939	Chillán, Chile	28 000	8.3
1948	Turkmenistan	110 000	7.3
1960	Agadir, Morocco	12 000	5.7
1960	Chile	5000	8.3
1963	Skopje, former Yugoslavia	1000	6.0
1964	Anchorage, Alaska	131	8.4
1970	Northern Peru	66 794	7.7
1970	Yunnan, China	10 000	7.7
1976	Guatemala City, Guatemala	23 000	7.5
1976	Tangshan, China	242 000	8.2
1977	Bucharest, Romania	1600	7.2
1978	Tabas, Iran	25 000	7.7
1980	Al Asnam, Algeria	4500	7.3
1985	Mexico	25 000	8.1
1988	Armenia	25 000	6.9
1989	San Francisco, USA	62	7.1
1990	North-west Iran	50 000	7.7
1995	Kobe, Japan	5500	7.2

LIVING WITH DISASTER *As long as people continue to live and build in communities in earthquake-prone areas, such as Los Angeles (right), there will be earthquake disasters. The day may come when earthquakes can be predicted. They will never be controlled.*

TSUNAMI: WAVE OF TERROR

Imagine a wall of water surging up the street, lifting cars and flinging them through flimsy buildings as it goes. Then imagine the buildings themselves being washed away, leaving behind little more than scattered piles of rubble.

At approximately midday GMT – 1 am local time – on April 1, 1946, a tremor in the fault system beneath the Aleutian Trench sent millions of tons of sediment cascading down its steep, unstable northern flank, producing a massive disturbance in the normally placid waters of the Pacific Ocean.

The tremor was recorded by seismographic stations throughout the world, including the Hawaiian Volcano Observatory, 2300 miles (3700 km) away. What they did not record was the deadly side effect that soon followed.

Twenty minutes after the earthquake, the lighthouse at Scotch Cap on Unimak Island, Alaska, about 95 miles (150 km) from the epicentre, ceased to transmit. The whole structure had been swept away by a 100 ft (30 m) wave that crashed down on it out of the night.

When a pebble is thrown into a pond, there is a splash, followed by ripples, which spread out in a concentric circular pattern to the pond's edge. The pond in this case was the Pacific Ocean, and the ripples were seismic sea waves travelling at over 450 mph (700 km/h) – a phenomenon that is often misleadingly called a 'tidal wave' (because it has nothing to do with the tides). In earthquake-prone Japan, where these occurrences are facts of life, they are called tsunamis – and this has become the accepted scientific term.

DESTRUCTION OF HILO ***A man (arrowed, top) watches helplessly as a tsunami surges towards him over a pier. Moments later, he was dead. The wave went on to demolish a large part of the town, and in particular its wooden buildings (right).***

Five hours after the submarine earthquake in the Aleutian Trench, Hawaii was beginning the day. In the northern town of Hilo, children were making their way to school; the Japanese community was preparing for work; businessmen were waiting at commuter train stations and bus-stops; in the harbour, dock-workers and sailors were already unloading.

The first indication that something was happening came at 6.50 am, when the water suddenly receded from the shoreline and the harbour basin, like a sudden outgoing tide. Loose boulders rattled in the violent backwash, which exposed a flank of glistening black volcanic rock and wet sandbanks. With no sea water to slow its flow, the local river increased its speed, pouring its waters into the empty river mouth. Many people saw this and waited to find out what would happen next. Others, more experienced, ran inland, shouting warnings. The date was not auspicious – April 1, All Fools' Day. In too many places, the cries of danger were ignored. The first wave, channeled in by the funnel shape of the bay and slowed by the shallowing of the shoreline, reared up in a foaming

WASHED UP ***The power of a tsunami is often demonstrated by the fact that boats are lifted bodily from their moorings and flung far inland, as in Alaska in 1964 (right).***

breaker and crashed through the harbour front across the lower areas of the city. At the river mouth, one span of the girdered railway bridge was wrenched from its piers and flung up the channel. Along Kamehameha Avenue, the town's major seafront road, buildings were uprooted from the seaward side and crushed into others on the landward side; all were smashed to rubble and splintered wood. The timber buildings of the Japanese district were flattened, except for a sturdily built bottling plant to which many inhabitants had fled on hearing the warning. In more well-to-do areas, people who heeded the alarm took refuge on the roofs of their substantial houses as the surging water blasted through the rooms below them, plucked their cars from their garages and carried them away. But this was only the first wave. Minutes later the second arrived, picking up the debris left by the first and flinging it far and wide. This second wave caused even more damage, sweeping away people who had left their places of safety to try to rescue others. A whole series of waves followed. The seventh was the biggest: it topped the palm trees on an island in the harbour and demolished the seafront hotel, which had withstood the earlier assaults. Fishing boats were driven ashore and capsized; railway lines were lifted and buckled; and wagons were hurled through the lower storeys of railway buildings. The business district of Hilo was reduced to a wasteland of scattered heaps of wood and rubble. It was 9 am before the last of the waves had passed. By then, 159 citizens of Hilo were reported dead or missing, and the survivors began to take stock of what had happened.

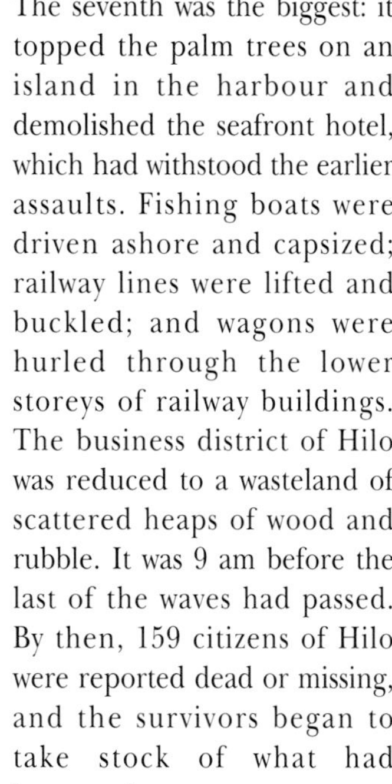

Under the Volcano *The landscape is quiet now, but in 1883 the volcanic edifice beneath these islands erupted in an explosion that sent tsunamis halfway round the world.*

What Causes a Tsunami?

Hilo had been the victim of a tsunami – a vast sea wave generated by some catastrophic event beneath the surface of the sea. The most common cause of this is an underwater earthquake, in which portions of the Earth's crust on either side of a fault jolt past each other. For a tsunami to occur, however, there must also be some vertical movement along the fault, capable of displacing huge amounts of sea water. Earthquakes observed on land rarely have a vertical displacement of more than a yard, which is probably too insignificant for such a devastating result on its own; under water, however, the side effects might include a landslide of loose sediment, shaken down the continental slope by the earthquake.

The Aleutian Trench, where the earthquake that generated the Hilo tsunami of 1946 originated, is 26 000 ft (8000 m) deep and 2000 miles (3200 km) long, and marks the line along which the floor of the Pacific Ocean curves down and cuts deep into the body of the Earth, beneath the continental masses of north-east Asia and north-west America. This is not a gentle process: it proceeds in a series of catastrophic jolts, freeing the rocks from the rock-jam that has held them rigid during the decades or centuries between each movement. It was one such jolt that occurred close to 1 am on April 1, 1946.

The eruption of an undersea volcano can also cause a tsunami. If a volcano has been erupting continuously for a long period of time, the magma chamber – the vast cavity within the Earth's crust that holds the molten material that is ejected – becomes

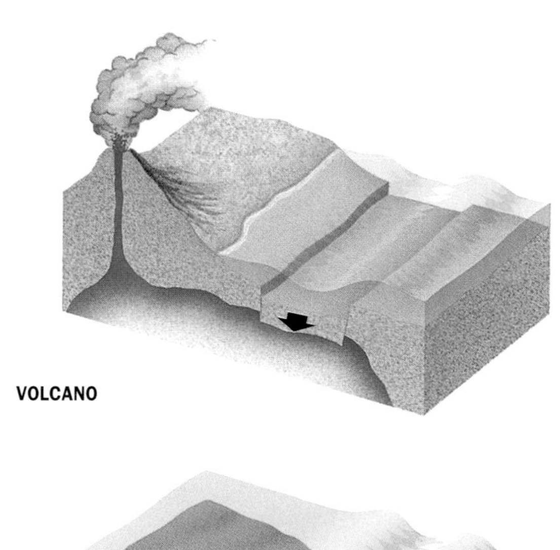

VOLCANO

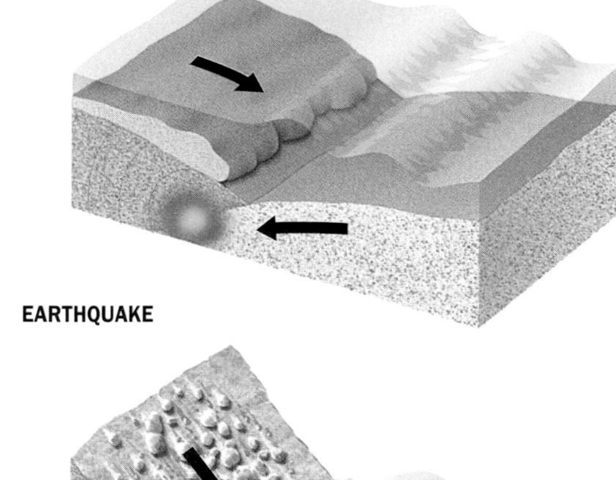

EARTHQUAKE

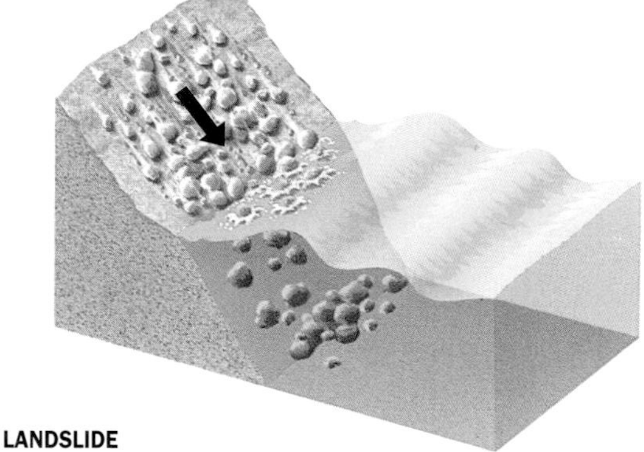

LANDSLIDE

Underwater Movement *Any undersea disturbance can cause a tsunami, particularly subsidence associated with a volcano (top), an earthquake along a subduction zone (middle), or a landslide falling into deep water (bottom).*

STRANDED VESSELS *Huge steamers were washed ashore on the islands of the East Indies by a tsunami generated by the eruption of Krakatau.*

depleted. The roof may then collapse, forming a crater or caldera, perhaps half a mile (1 km) in diameter. Tremendous volumes of water fall into this undersea crater in a very short space of time, causing a tsunami. One of the most devastating tsunamis ever known was generated in 1883 by the collapse of the volcano Krakatau in what is now Indonesia; a total of 36 000 people were killed.

Thirdly, tsunamis are also caused by a landslide or avalanche falling directly into the sea. For example, in the spring of 1958, heavy rain and a rapid thaw weakened the rocks at the head of an inlet in the Gulf of Alaska. An earthquake then shook the mountainside loose and dropped it into the inlet with such force that it sent up a 100 ft (30 m) wave. The wave surged back and forth along the inlet, flinging itself up the rocky sides as it swept around corners, stripping away soil and trees that were growing more than 1700 ft (510 m) above sea level.

A seiche, which is not unlike a tsunami, is a wave motion set up in water by the arrival of earthquake tremors. Seiches usually coincide with the earthquake vibrations, whereas a tsunami arrives later. The Lisbon earthquake of 1755, which took place at 9.30 am, generated seiches in the English Channel half an hour later; the tsunami associated with the earthquake did not reach the English Channel until well after noon.

Once a tsunami is released, it spreads across the ocean very much like an ordinary wave created by the wind: the greatest difference is one of scale. Most sea waves travel at a few miles per hour, but a tsunami in the open ocean will travel at the speed of a jet plane. At such speeds, the wave length is very long – more than 60 miles (about 100 km) – but the height is only about 3 ft (1 m), which means that a tsunami is not easily detectable in the general movement of sea waves and can pass beneath ships unnoticed. The Krakatau tsunami passed unremarked through the world's busiest shipping lanes. It is only when the seabed shallows or islands and continents stand in the way that the full destructive potential of a tsunami is unleashed.

THE GRAND BANKS UNDERWATER LANDSLIDE

The submarine landslides that cause tsunamis can be difficult to study, since they are so unpredictable and out of sight. In 1929, however, an undersea earthquake triggered a landslide on the Grand Banks near Newfoundland. The mass of loosened rock slid down the continental slope, destroying undersea telegraph cables as it went. One by one the telegraph cables ceased to transmit; and by looking at the times of the interrupted signals, scientists could work out the speed at which the landslide moved. The mass covered a distance of nearly 310 miles (500 km) at speeds of about 60 mph (100 km/h).

HOW A TSUNAMI TRAVELS

It is not actually the water itself in a tsunami – any more than it is the ground in an earthquake – that travels great distances at high speed; the water simply transmits the disturbance. Like water particles in an ordinary wave, the particles in a tsunami move in a circular pattern, up

CITYSCAPE *The great Lisbon earthquake of 1755 was accompanied by devastating tsunami, shown on this contemporary engraving.*

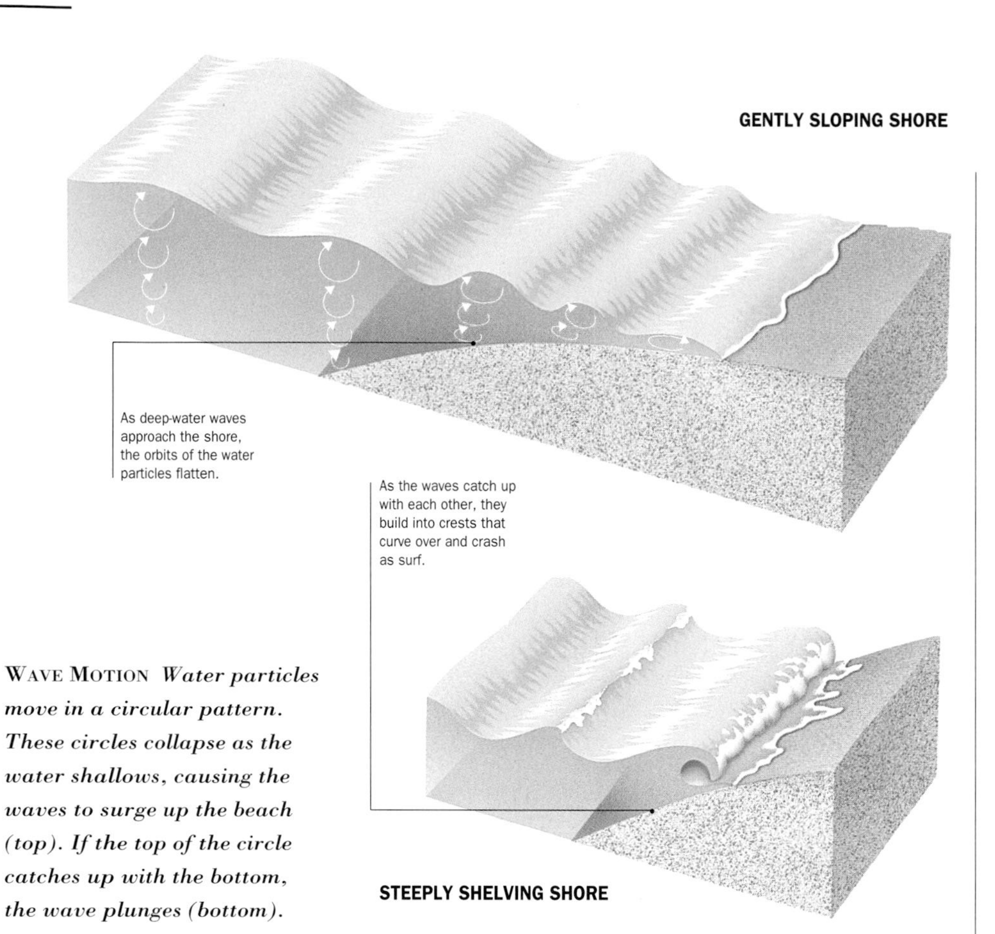

WAVE MOTION *Water particles move in a circular pattern. These circles collapse as the water shallows, causing the waves to surge up the beach (top). If the top of the circle catches up with the bottom, the wave plunges (bottom).*

and over, down and back – eventually returning to their first position. This circular motion is greatest at the surface but dies away at depths where friction with the seabed slows it down. In the open ocean, where the seabed can be several miles below the surface, there is very little to obstruct the passage of a tsunami. In shallower parts of the ocean, however – for example, as the wave approaches the shore – the water nearer the surface begins to pull away from the slower-moving deeper water, and to accumulate in massive waves that topple forwards and break. In the case of the tsunami, the individual wave crests catch up with one another, and pile together in a single towering wave that hits the shore with devastating effect.

CRASHING ASHORE

Hawaii is particularly vulnerable to this process. The island chain is situated in the middle of the Pacific Ocean, an ocean fringed by active ocean trenches, all of which are potential undersea earthquake sites. What is more, Hawaii is made of shallow volcanoes, shaped like medieval shields lying on the ground, that spread over a wide area beneath the ocean surface and afford very gently shelving approaches to the coastline. These are just the conditions that build up tsunamis to their destructive height. Indeed, the very word tsunami, which literally means 'harbour wave', emphasises the fact that the phenomenon is only significant, or dangerous, in shallow waters.

By contrast, the other Pacific islands do not fare quite so badly. Rocky outcrops rising sheer from the ocean floor, such as Pitcairn Island, or steep coral atolls that have developed from the subsiding stumps of ancient volcanoes, such as the islands of Kiribati, are not surrounded by shallow waters; and tsunamis tend to pass them by without causing much damage.

Japan is another vulnerable area – a volcanic island chain near the junction of three oceanic trenches, all constantly active. Tsunamis generated here attack the shore with little or no warning. Those that occur in the Sea of Japan, between the Japanese island arc and the mainland of Asia, can spread out and hit a number of coasts, bouncing back and reinforcing one another, and producing unforeseeable patterns of destruction.

The shape of a coastline has a great effect on the characteristics of a tsunami as it hits. Narrowing inlets and bays, like those of Hilo Bay in Hawaii, will funnel the waves inwards, building them even higher. This is part of the reason that the tsunamis of 1946 were so destructive in Hilo. A normal

THE CRUEL SEA *The horrifying power of a tsunami is suggested by this photograph of a winter wave plunging on to the beaches of Hawaii.*

STRANGE NOISES

Mariners in the vicinity of an undersea earthquake often report hearing explosions, like distant gunfire. These are sonic booms generated by the earthquake's shock waves disturbing the surface of the water; these P-waves move at about 12 500 mph (20 000 km/h), which is far greater than the speed of sound.

COASTAL INDUSTRIES *Fishing communities are usually badly hit by tsunamis, as seen in these photographs of the destruction on Okushiri Island after the earthquake of 1993.*

wind-generated wave train approaching a beach at an angle will be bent, or refracted around, so that it hits the beach full on. Similarly, a normal wave train hitting a headland will be refracted around so that the waves hit the headland from either side – which explains why headlands erode. The same can be true of tsunamis. On July 12, 1993, a tsunami generated in the Sea of Japan swept down on the city of Okushiri, built on a promontory on Okushiri Island. The waves curved in from each side; as the shoreline grew shallower, they rose to heights of 35 ft (10 m) before plunging down on both beaches of the city, leaving 240 people dead.

Normal waves coming ashore usually form breakers: spilling breakers, in which the crest loses its stability and breaks up into foamy turbulence; surging breakers, in which the whole mass of water rushes up the beach; or plunging breakers, in which the front of the wave rises to form a vertical wall, the crest curving over and tumbling in front of it. The popular view of a tsunami is of a plunging breaker thundering down on to a populated shoreline. However, the physics of wave formation on such a scale make gigantic plunging breakers unlikely, and when tsunamis do form breakers, they are more likely to be of the spilling or surging kind. More often, they do not form breakers at all, and the result is merely a sudden rise in sea level – flat-surfaced, with no discernible waves.

In certain circumstances, however, a steep-fronted wave may form. A tsunami funnelled in through a narrowing inlet rushes so quickly that it piles up its waters and pushes them forwards in a turbulent front, as was observed at Hilo in 1946. This is the same process that forms a tidal bore, when the rise in sea level caused by a tide is restricted in a narrowing estuary: the result is a surge of water upriver, as can be seen on the lower Seine in France, the Solway Firth between Scotland and England, and the River Severn in England.

TSUNAMIS OF THE PAST

You would expect a phenomenon as destructive as a tsunami to leave behind some evidence of its passage in the distant geological past, or even in the more recent historical past; and rock strata and archaeological sites do provide such data.

Dinosaurs died out some 65 million years ago. According to a current theory, a gigantic meteorite slammed into the Earth and caused so much environmental disruption that many types of animal became extinct. Although there is an eroded crater of the right age in Yucatan, there is no evidence of a causal link between it and the extinction. If the meteorite had landed in the sea, on the other hand, all the evidence would long ago have been effaced. It is almost impossible to imagine the scale of a tsunami generated by a vast meteorite landing in the sea. In recent years, however, geologists have uncovered very strange rock formations dating back 65 million years, in Texas, Alabama, Haiti and Mexico. These rocks consist of a layer of molten rock globules, overlaid by masses of what may well be deep-sea rocks in the form of huge boulders, as well as shallow-water rocks and even

HOW DOES HOLLYWOOD TREAT A TSUNAMI?

On movie locations, tsunamis are usually depicted by placing a model boat in the path of a plunging breaker and filming the result in slow motion – scarcely a realistic portrayal of a tsunami in the open water. *Fair Wind to Java* (1953) used this technique to re-create the Krakatau tsunami – destroying the villain's ship, but not the hero's, in the best Hollywood style. On other occasions, tsunamis are simulated with wave machines in a miniature tank. When this was done to capsize the liner in *The Poseidon Adventure* (1972), the script had a member of the crew acknowledge that the wave was building up in shallows, giving some reason for a tsunami to form a spilling breaker in the open ocean. Both techniques were used in *Krakatoa, East of Java* (1968). Krakatoa – an alternative spelling of Krakatau – is actually west of Java, but it sounds less romantic.

The most striking cinematic tsunami occurs in *When Worlds Collide* (1951), as a rogue planet passing close to the Earth sets off a string of natural disasters, including the inundation of Manhattan by giant waves. The effects sequences won an Academy award, and actual footage has been repeated in many films since, including *Atlantis, The Lost Continent* (1961) and *The Seven Faces of Dr Lao* (1964). Much of this was achieved by using dump tanks that spilled thousands of gallons of water on to miniature sets. In the Manhattan scene, the water swirled around blocks representing the city's buildings. Film of the buildings themselves was then superimposed on the result.

Other films featuring tsunamis include *Meteor* (1979), in which Hong Kong was inundated by the same photographic process that destroyed Manhattan in 1951, and *Tidal Wave* (1975), an Americanised version of an earlier Japanese disaster film, *Submersion of Japan*. *Clash of the Titans* (1981) and *The Abyss Special Edition* (1993) also featured giant waves, generated by monsters and aliens rather than by normal geological processes.

When Dinosaurs Ruled the Earth (1970) did show the withdrawal of the sea from the shoreline as the first sign of an approaching tsunami, as in Hilo. Unfortunately, this was followed by less imaginative, slow-motion shots of a plunging breaker and familiar stock footage from *When Worlds Collide*.

'Krakatoa, East of Java'
A miniature ship is struck by waves in a tank in a film-maker's studio.

'When Worlds Collide'
Manhattan is engulfed by giant waves in a film of 1951.

fragments of fossilised trees. Above all this, there is a layer of sandstone lying in beds that were probably formed by very violent and turbulent currents.

Analysis suggests that a meteorite impact in the ocean vaporised part of the Earth's crust, which was then scattered about in molten globules. Massive tsunamis would have followed. Between 1500 and 3000 ft (500-1000 m) high, they would have ripped up the ocean bed and flung it ashore, mixing it with shallow-water muds and coastal vegetation. Subsequent tsunamis, it is said, then swept back and forth for a long time before the water finally settled. This scenario is consistent with the latest theory that such a meteorite impact actually took place in the Gulf of Mexico. Further research into the rock formations of the area, carried out in 1994, suggested however, that more conventional sea currents, volcanic eruptions and weathering effects could just as readily account for the character of the rock beds.

There is also evidence to suggest that a prehistoric tsunami occurred in the south-eastern Pacific 105 000 years ago, after an enormous undersea landslip off Hawaii. This was probably produced by the continuous growth of Hawaii's volcanoes and may have measured 15 miles (25 km), parallel to the coast, with a thickness of 6 miles (10 km). The discovery of this evidence on the sea floor spurred geologists to look for other signs of tsunamis – which they found in the strange distribution of sand dunes in south-eastern Australia. All the parts of the coastline facing towards Hawaii, which were not sheltered by island chains or by the Great

PORT DAMAGE *An earthquake and a tsunami in Alaska in 1964 caused scenes of devastation such as this.*

Barrier Reef, had been scoured clear of dunes at about that date; and rocks weighing up to 20 tons had been plucked from their cliffsides. For this to have happened, tsunamis towering to heights of between 65 and 160 ft (20-50 m) must have descended upon them. On the sheltered south-western sides of the headlands, however, the sand dunes are intact.

Before about 900 BC, the major centres of Peruvian civilisation lay along the Pacific coast. Thereafter, however, the coastal settlements seem to have given way to mountain cultures, of which the first was the religious centre of Chavín de Huantar, perched 10 000 ft (3000 m) up in the Andes. The city contained a proliferation of artistic motifs that were obviously drawn from the earlier coastal culture. Could a tsunami have wiped out the coastal villages, whose survivors then moved to the mountains? Such a theory is largely corroborated by the discovery in Lima during the late 1980s of damaged buildings, layered over with silt. The archaeologists performing the excavation attributed their findings to the effects of a giant tsunami and its backwash.

Tsunamis that have taken place in historical times are of course better documented. Since the rise of civilisation in the Middle East, there have been records of destructive waves sweeping the volcanic and earthquake-prone Mediterranean Sea; populated areas of the Pacific, such as Hawaii and Japan, have always been prey to tsunami; and even the comparatively stable Atlantic Ocean has had its own tsunami disasters – the best-known being the wave that swept Lisbon following the earthquake in 1755, which was recorded as far south as Morocco and as far north as the English Channel.

The most thoroughly studied of the historical tsunamis followed the eruption of Krakatau in the East Indies in 1883. This uninhabited volcanic island in the Sunda Strait between Java and Sumatra had been erupting intermittently for several months. Then, on August 27, came the climactic eruption that sent 1 1/2 cu miles (6 km^3) of ash 30 miles (50 km) into the air in a blast that was heard in Australia. The seabed collapsed into the empty magma chamber below, swallowing almost the entire island and forming a caldera 5 miles (8 km) wide and 650 ft (200 m) below sea level. Following the landslide, a tsunami swept along the coasts of Java and Sumatra, where it seems to have exceeded 100 ft (30 m) in height and obliterated 165 villages. Some 5000 miles (8000 km) across the Indian Ocean in South Africa, it registered a height of about 1 ft (30 cm), and it even produced a 2 in (5 cm) surge in the English Channel, 11 000 miles (18 000 km) from its place of origin. Some 36 000 people had perished in the disaster.

PREDICTION AND WARNING

The 1946 disaster in Hilo showed the world that some means of monitoring tsunamis was essential. It had been known for a long time that undersea earthquakes were

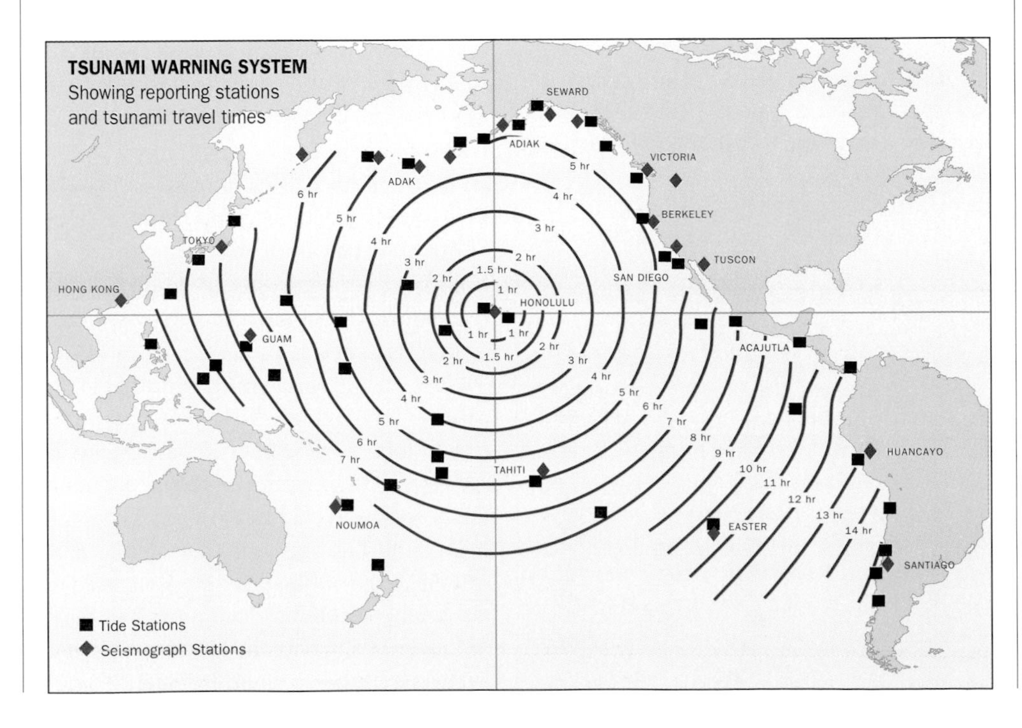

TSUNAMI WARNING *Charts can be used to predict the arrival of a tsunami, but nothing can be done to stop one.*

responsible, and seismographs capable of detecting such earthquakes had been installed in the world's observatories for several decades. Some attempt at predicting tsunamis had been made after the Hawaiian Volcano Observatory was set up in 1912, but there had been a number of false alarms and the system had little credibility by the time of the Hilo disaster.

By 1949, however, a reliable Pacific-wide system was in place. The Tsunami Warning System (TWS) was based in Hawaii's Honolulu Observatory, and linked to observatories in Sitka, Alaska, and Tucson, Arizona, as well as to tide-monitoring stations throughout the Pacific. A single seismograph can estimate the distance, but not the direction of an earthquake. Three seismographs, however, can pool their information and pinpoint the earthquake's focus. Should an earthquake measuring more than 7 on the Richter scale be observed in, or close to, the Pacific Ocean, then a Tsunami Watch is instigated. This involves monitoring the tidal gauges at the five stations nearest to the earthquake's epicentre for any abnormalities in sea level, and alerting the emergency agencies. In Hawaii, public warnings are issued on all the broadcast channels, and low-lying coastal areas are evacuated; coastguards and any other boat-owners are advised to put to sea and to make for waters deep enough to ensure their vessels' safety. The approximate times of arrival of the tsunami in Hawaii can be estimated by using a chart published by the National Oceanic and Atmospheric Administration.

Recently, Japan has led the field in tsunami forecasting. The issue is even more critical there than in Hawaii, since tsunamis are generated by earthquakes that originate in the ocean trenches close by, with very little time lag between earthquake and tsunami impact. The Japanese Meteorological Agency (JMA) has had warning systems installed for many years, but they have not been responsive enough. In May 1983, an offshore earthquake generated a tsunami that flooded the Akita coast of north-west Japan killing 104 people – the tsunami warning did not come until six minutes after the tsunami had struck. In 1994, the JMA installed 150 seismometers throughout the country, designed to send instant signals to the headquarters in Tokyo. Factors such as the magnitude of the earthquake, its depth of focus, and the depth of the sea can be instantly analysed by computer, and warning captions then broadcast on all of Japan's television channels. The Real Time Seismology System, as it is known, was triggered on the night of October 4, 1994. An earthquake measuring 7.9 on the Richter scale occurred on the ocean bed off the east coast of Hokkaido, 590 miles (950 km) from Tokyo. The warning was flashed on television screens within four minutes. When the tsunami struck, about an hour later, there were 200 minor injuries, but no fatalities. The system had worked. Such systems are improving all the time; modern tidal gauges are sensitive enough to register the slightest changes in water depth.

MAJOR TSUNAMIS

Date	Origin	Effects	Death toll
June 7, 1692	*Puerto Rico Trench, Caribbean*	*Port Royal, Jamaica permanently submerged*	*2000*
November 1, 1755	*Atlantic Ocean*	*Lisbon destroyed*	*60 000*
August 8, 1868	*Peru-Chile Trench*	*Ships washed several miles inland Town of Arica destroyed*	*10-15 000*
August 27, 1883	*Krakatau*	*Devastation in East Indies*	*36 000*
June 15, 1896	*Japan Trench*	*Swept the east coast of Japan, with waves of 100 ft (30.5 m) at Yoshihimama*	*27 122*
December 28,1908	*Sicily*	*East coast of Sicily, including Messina, and toe of Italy badly damaged*	*58 000 (including earthquake victims)*
March 3, 1933	*Japan Trench*	*9000 houses and 8000 ships destroyed in Sanriku district, Honshu*	*3000*
April 1, 1946	*Aleutian Trench*	*Damage to Alaska and Hawaii*	*159*
May 22, 1960	*South-central Chile*	*Coinciding with a week of earthquakes Damage to Chile and Hawaii*	*1500 (61 in Hawaii)*
March 27, 1964	*Anchorage, Alaska*	*Severe damage to south coast of Alaska*	*115*
August 23, 1976	*Celebes Sea*	*South-west Philippines struck, devastating Alicia, Pagadian, Cotabato and Davao*	*8000*
December 12, 1992	*Flores Island, Indonesia*	*Damage to South-central Indonesia*	*1000*

NIGHT LIGHTS

Some fortunate survivors of nocturnal tsunamis have reported that the wave actually glows as it approaches. This is probably due to the presence of luminous algae disturbed by the original tremors from the earthquake. These then swarm to the surface of the water, where they are concentrated in the spilling chaos of the wave's crest. There are many such marine organisms that emit light when disturbed. The most common are the floating microscopic algae, such as the dinoflagellate *Noctiluca miliaris* (with two whip-like flagella). These are visible as they cause the wakes of any ships that pass through swarms of them when they are in bloom to glow. The effect of millions of these microscopic light sources is to produce an even glow when the water is disturbed.

2 The Shifting Surface

Avalanche *A fluffy cloud of snow slips and tumbles down a Himalayan mountainside.*

Far from being the symbols of eternity that they may seem, many of the world's mountains are quite young. Built by the titanic collision of the continents, they are eventually ground down by the forces of erosion — and in particular, by running water and the wind. The forces of gravity then take over, transporting the particles to the sea. Gravity ensures that one hundred billion tons of rock, whittled away from the Earth's surface, are carried down into the oceans each year. Sometimes they creep slowly downhill, but at other times a sudden rockslide might cause a whole mountainside of unstable rock to career into the valley.

Landslide *An Alpine hillside collapses into a jumble of rocks.*

LANDSLIDES AND ROCK MOVEMENTS

Throughout the world, wherever new mountain ranges are being formed, rocks are gradually being folded and forced up to great heights. Balancing this is the work of gravity which, from time to time, brings them tumbling down again.

The slate quarry cut into the forested mountainside was a source of income for many of the villagers of Elm in the north-western Swiss Alps. It also provided a fatal weak spot that led to the obliteration of the village. The quarry was halfway up the mountain, some 1100 ft (335 m) beneath the crest of the Plattenbergkopf ridge which separated the valley from its neighbours. The first cracks – telltale signs of a mountainside under stress – appeared on the steep slopes overhanging the quarry. They formed slowly, over a period of 18 months, but when disaster struck in 1881 it was without warning, swift and deadly.

One particular crack had cut a diagonal gash, 30 ft (9 m) deep, that ran from the crest of the ridge to the edge of the quarry, exposing the very structure of the rock face to the late summer rains. On a September afternoon, a slice of mountain suddenly split away and fell downhill, disintegrating into a mass of rubble only paces away from the Elm village inn. It was a short-lived reprieve; just over a quarter of an hour later, a second, bigger landslide crashed down, crushing the inn and several neighbouring houses. Four minutes after that, the entire mountainside above the quarry collapsed and careered down the steep slopes as a single mass of land, the forest still clinging to it, trees quivering like corn in a wind-blown field. It poured onto the quarry floor with such impact that it burst and exploded into the valley below.

The Elm landslide consumed everything in its path, carrying some rocks the size of houses, pulverising others with its sheer bulk and speed. As it collided with the opposite side of the valley it reared 3000 ft (900 m) up the slopes and then swirled around to thunder along the fertile valley floor. It took less than a minute for the final devastating collapse of the mountainside to bury the village and 115 of its inhabitants beneath 30-60 ft (9-18 m) of shattered rock and debris.

DECONSTRUCTING A MOUNTAIN *The mountains of southern Iran, shown in this false-colour satellite image, were created by the collision of the Iranian and the Arabian tectonic plates. They are constantly being eroded, however.*

THE FORCE OF GRAVITY

Such dramatic land movements are most likely to occur in unstable mountain ranges like the Himalayas, the Alps, the Andes and the North American Coast Ranges, where mountain-building processes are still active. The rocky plates of the Earth's

LIKE A WATERFALL *Rocks tumble down a slope in Zion National Park, Utah (right), churning and flowing almost as if they were liquid.*

DANGER: CONSTRUCTION IN PROGRESS

Landslides and rock falls, earthquakes and volcanic eruptions, all occur at vulnerable and unstable points in the Earth's surface, where the process of mountain-building is still active. The outer crust of the Earth is as thin in relation to the planet as a whole as the skin of an apple in relation to its flesh. It is made up of a number of massive rigid sections or plates, which float like rafts on the semi-solid rocks beneath, moved by convection currents generated by heat from the Earth's core. It is where these plates meet and are forced against – or pull away from – each other that ocean trenches, rift valleys and certain mountain ranges are formed and the Earth's worst pressure points occur.

There are two types of plate: dense, heavy ocean plates and lighter continental plates. Both occur and meet around the rim of the Pacific Ocean. As massive forces push them against each other, the ocean plate gives way, slides beneath the continental plates and is destroyed. At the point where it sinks, or is 'subducted', an ocean trench forms – like the Peru-Chile trench in the eastern Pacific – and on the continental margin, hills and mountains are rumpled up or 'folded' into being. Over geological time, sediments brought down by rivers from the adjacent landmass and from the ocean itself build up. The weight of accumulated layers and pressure from the continuing subduction process compresses the sediments into sedimentary rocks, and at the margins of the continental plate these may be crumpled into hills. Sometimes, the pressure causes the rocks to shear and split into wedges that pile up against one another, as along the western coast of South America.

Unstable Area ***Fold mountains are constantly being pushed up, continually providing fresh slopes and precariously high rock outcrops that are vulnerable to gravity.***

Other sedimentary rocks on the ocean plate are dragged deep below the continental mass, where they are subjected to such pressures and temperatures that their structure changes and they become what geologists call metamorphic rocks – from the Greek *meta*, involving change, and *morphe*, 'shape'.

Missing Rock ***The ragged edges of Sheep Mountain, Wyoming, are all that remain of the vast volumes of rock that have eroded and fallen away at some time in the past.***

The ocean plate sinks still farther and eventually melts as it approaches the Earth's molten core. The liquid rock is now lighter than the rocks above and rises through them. When it reaches rocks of similar density, it splays out horizontally in mushroom-shaped blobs known as plutons and becomes the magma chamber of a volcano, or wells up to the surface and solidifies. These rocks born from fire are described as igneous (from the Latin *ignis*, 'fire'), and like metamorphic rocks, sometimes only become exposed at the surface if upper layers of sedimentary rock are worn away.

By such processes, successive ridges of land are built up seawards, at the edges of continental plates. Away from the plate margin, the pressures may still be felt, splitting the continental mass into chunks that are forced over one another.

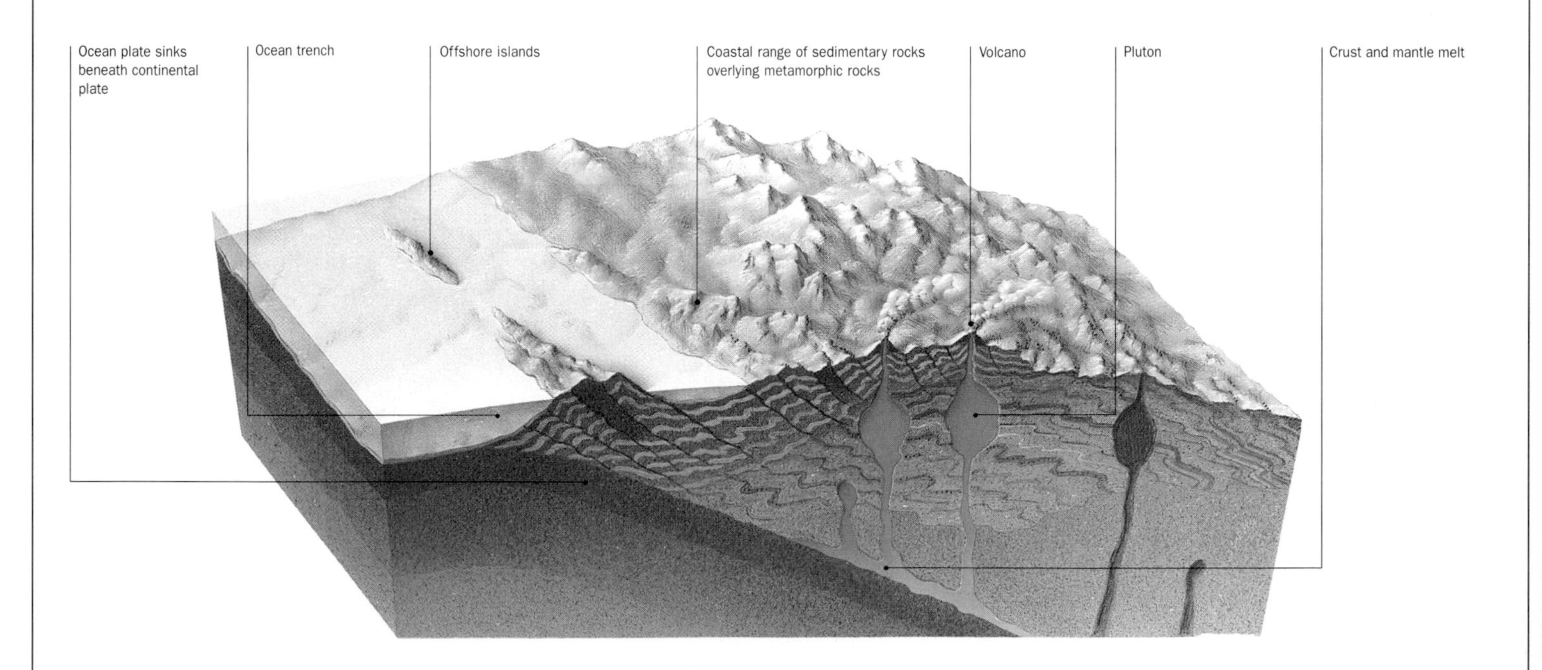

Before and After *The peaceful Peruvian town of Yungay (left) was transformed in a few minutes in 1970 into the waste of rubble (right), as the mountainside fell upon it.*

outer shell push against, and ride over, one another forcing the landmasses at their edges to buckle and fold into mountains; the land at these points is constantly rising and the slopes steepening, making the rocks very unstable. Simultaneously, the destabilising and destructive forces of erosion and gravity are at work – a process often spun out over millennia but sometimes striking with devastating suddenness.

One of the greatest landslide disasters in history gave no warning at all, and took only seven minutes to wipe out the population of several towns and villages.

Sliding Soils of South America

At 21 830 ft (6654 m), Nevados Huascarán in the Cordillera Blanca is the second highest peak in South America and regarded as the most unstable mountain in Peru. And yet those who lived in the shadow of the mountain in the valley of the Río Shacsha felt no sense of impending disaster on January 10, 1962 – until a thunderous crash shattered their afternoon siestas, a crash so powerful that it resounded through the mountains hundreds of miles away. A 3-million-ton block of ice had split from a melting glacier at the valley head and exploded into the gorge below. Here it turned into a wave of destruction, scouring the gorge of soil and forest and wrenching rocks from the very fabric of the mountain. The inhabitants of Pacucco, Yanamachito, Huaraschucho, Uchucoto, Chiquibamba and Calla villages could only watch in horror as the torrent of ice, mud and rocky debris roared towards them moments before they were engulfed. Then, suddenly, the landslide came to a halt, its energy expended. In seven minutes, it had travelled 10 miles (16 km) leaving a valley choked with debris and more than 4000 people dead.

The event was almost insignificant when compared with what happened eight years later in the same region. On May 31, 1970, an earthquake measuring 7.7 on the Richter scale shook the depths of the Peru-Chile trench in the eastern Pacific Ocean. Nearly 80 miles (125 km) away, the side of Huascarán mountain shuddered and collapsed. About 3500 million cu ft (100 million m^3) of rock plummeted 2000 ft (600 m) down a near-vertical cliff face into the densely populated valleys below, ripping out a glacier and its moraine en route. Now transformed into a 330 ft (100 m) deep torrent of rocks and mud, the landslide roared 9 miles (14 km) down the valley at more than 175 mph (280 km/h), flinging house-sized boulders into the dust-filled air. Its approach was heralded by explosive sound, ground and buildings trembling in the shock wave of air that preceded it.

The monster devoured the waters of the river Shacsha and became a mud flow with a prow a mile (1.6 km) wide. Towards the mouth of the valley the mud slide split. In one direction lay the town of Yungay, whose

USEFUL AVALANCHES

Before the dissolution of the USSR in 1990, there were plans deliberately to create rock avalanches that would block valleys in the Tein Shan mountains, near the border with the People's Republic of China. Reservoirs would form behind the resulting dams, to provide hydroelectric power and a steady supply of water for irrigation.

18 000 people were impotent before the mass of debris churning towards them. A hundred or so fled to the high spot of the cemetery hill and survived. One of them described the advancing mud slide as a huge ocean breaker, more than 260 ft (80 m) high, with a continuous and deafening roar. In a few minutes, the town had ceased to exist, buried under 30 ft (9 m) of mud.

The other branch of the flow swept into the wide Rio Santo valley, obliterating the towns of Ranrahica and Matacoto. Then, saturated and swelled by river water, it cruised at 15 mph (25 km/h) for a further 30 miles (50 km) before finally stopping. Geological evidence reveals that an even greater landslide hit the same area in prehistoric times.

There are many different forms of landslide, all of which geologists refer to as 'mass wasting' – because large masses of soil and surface rocks are involved. At the opposite extreme to the Peruvian disasters is soil creep, which moves almost imperceptibly down a hillside at a rate of less than 3 ½ in (9 cm) a year. The most noticeable effects of soil creep include a tilting telegraph pole, cracks in a hillside road, a leaning wall, or a tree with distinct curvature of the trunk – as its natural upright growth battles against the downward pull of the slipping soil. Soil will be thin on the ground at the upper end of a field and piled against the walls at the bottom.

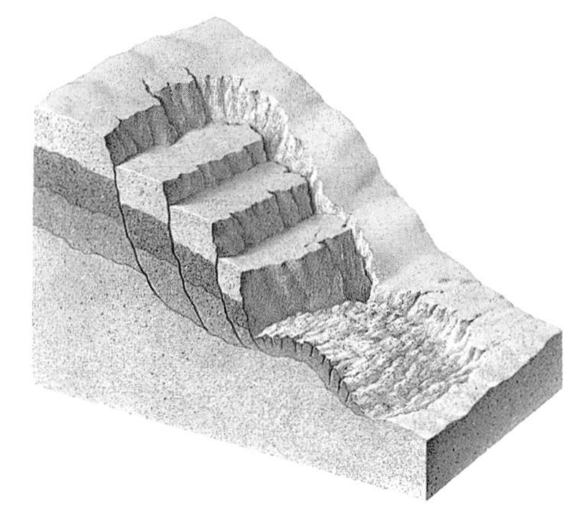

SLIPPING AND SLIDING *Once the soil has broken away in coherent, horseshoe-shaped slabs, it soon turns into a chaotic jumble and flows down the hillside in a great tongue.*

EARTH MOVEMENTS

Creep starts with soil being loosened by rainfall, frost action or the passage of animals and then being displaced farther downhill. Because it is bound by the roots of vegetation, it moves in slabs rather than as individual particles and settles into step-like terraces or mini-terraces that follow the contours of the hill. Soil creep is thus the movement of a coherent mass of surface material; if land slips downhill as a jumble of fragments, it is known as a debris flow, which moves a little faster, but not much.

On a large scale, soil creep can create problems. As fast as engineers sank the foundations for the Grand Coulee dam on the Columbia River in North America, water-logged silt from the surrounding site slipped into the excavations. The problem was solved by pumping a refrigerant into the silt to freeze and stabilise it until the

SOIL CREEP *The gradual movement of soil particles down a hill wrinkles the surface into a series of shelves called terracettes.*

foundations of the dam were completed.

The implications of mass wasting become more serious with the phenomenon known as rotational slip or slump. On a steep slope, a mass of weak rock such as clay can give way and slump downwards under the pressure of its own weight, spreading out at its base. This sudden slump produces a concave hollow in the slope shaped rather like a playground slide – almost vertical at the top and curving gently towards the horizontal at the bottom. As the rock slips down this concave slope (known as the sole), it tilts back like a body slumping into an armchair. An initial rotational slump at the top of a slope immediately puts the pressure on farther down, triggering a whole series of slumps like stepped hollows.

FLOATING TOWNS

Another form of mass wasting results in whole areas even of relatively flat land coming adrift. If you press your fingers into a lump of clay, they leave their imprint but the clay remains in one solid piece, its fine particles held together by the water contained within its pores. If the water content becomes too great, however, the slightest disturbance can cause the particles to fall apart and the clay to liquefy.

This happened on a grand scale with the Alaska earthquake of Good Friday 1964, which agitated the clay bedrock beneath the town of Anchorage and turned it into a sea of mud. The town was built on the gravelly crust that lay over the clay, on a gentle slope running down to the sea. When the clay liquefied, the crust, complete with its cargo of buildings, slid over it like an ice flow. Great slabs of the harbourside crashed into the sea. Buildings farther up the slope slid down in turn, and within little more than three minutes, most of the town had collapsed into the bay.

A similar fate befell the town of St Jean Vianney, situated on flat land in the St Lawrence valley in North America. In 1971, its inhabitants watched in amazement as their homes sailed sedately past them like ships in a harbour. The buildings collapsed and 31 people were killed. Here, as in other parts of the St Lawrence valley, a thick layer of fairly solid soil lay over slippery clay, causing the upper layers to slide – a phenomenon

QUICK CLAY *The soil, on which Anchorage was built, turned into a liquid during the earthquake of 1964 (above and top right). Buildings and roads slid away, lubricated by this mass movement.*

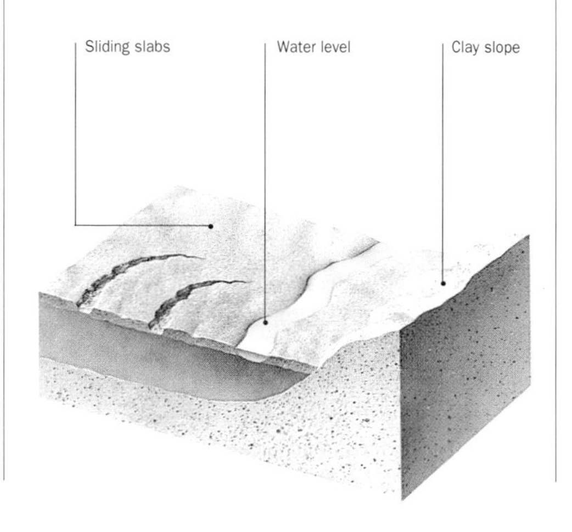

EARTH FLOW *Rather like mud flows, although less fluid, earth flows tend to be confined to downward slopes of weathered clay or shale.*

known as earth flow. A similar process happened in the Göta valley in Sweden in 1953 – Scandinavia being particularly prone to landslides.

Layers of rock of any kind, not just clay, may slide over each other if the bedding planes (where the different layers are sandwiched together) or other strands of weakness run parallel to the slope of the land. An example was the southern slopes of the 5164 ft (1574 m) Rossberg peak in Switzerland which dipped in line with the bedding planes. In 1806, the upper levels of rock disengaged from those beneath and hurtled into the valley below, burying the town of Goldau and its 457 villagers. It usually takes more than one type of mass wasting to wreak serious damage, however.

Rock Slide *A rock slide can result in a broad fan of broken rock at the bottom of the slope, as in this photograph of a cliff face in the Alps.*

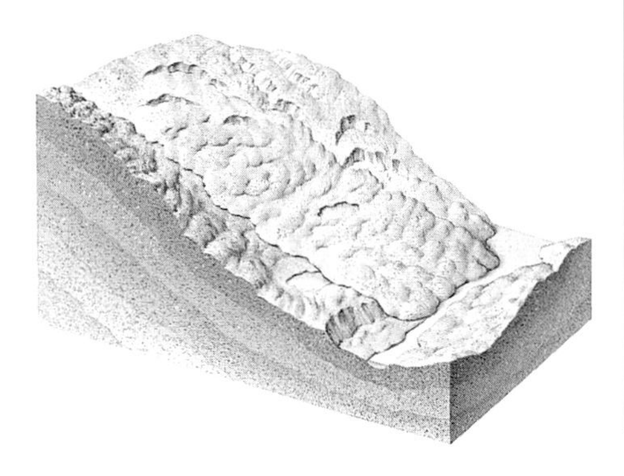

Debris Flow *Over the years, a debris flow can bring a whole mountainside down into the valley, as in the Gros Ventre valley, Wyoming.*

What begins as rotational slump may turn into debris flow as the slabs of rock splay outwards at their base and break up. The neat, horseshoe-shaped hollow of the textbook slump then degenerates into a mass of jumbled debris, slithering down the entire slope. The probability of rock slide increases if the base of a slope is weakened by the undercutting of wave or river erosion, or by a road, putting extra pressure on the upper slopes. The catalogue of events that began in 1908 in the Gros Ventre valley in Wyoming, and culminated in part of a town being washed away, was the result of several different factors.

Combined Forces

Heavy winter rain had saturated the southern slopes of the valley, causing a debris flow of surface soil and rock to begin tumbling downhill in May 1908. There was probably an additional element of rock

slide along an internal layer of weakness, and by 1911 a vast hummocky mass of soil and rock had accumulated on the valley floor. It was crosshatched with cracks and harboured ponds in shallow depressions. The valley road was wrecked and efforts to keep it open had to be abandoned. Most spectacularly, the pile of debris had dammed the river, so that a substantial lake, 5 miles (8 km) long and around 230 ft (70 m) deep, formed behind. It was not until 1927 that the lake, swollen by the waters of the spring thaw, spilled over the dam of debris. The loosely packed mud and rocks crumbled with the force of the water and the lake emptied rapidly. The suddenly released waters cut a 50 ft (15 m) deep channel through the material that had choked the valley and swept downstream to hit the town of Kelly.

In a rock fall, rocks bounce off cliff

ROLLS AND FREE-FALLS *A rock avalanche (far left) sends rocks rolling downhill, while in a rock fall (left) the rocks fall free.*

faces and fly through the air, and may mark the start of an avalanche. Once a rock avalanche has gathered momentum, it pours down a slope at speeds of up to 170 mph (275 km/h), like a raging torrent of water, swirling up and eddying around obstacles and finally settling in pool-like hollows. Air is squeezed beneath the moving debris, and steam is produced by water evaporating with the friction; both act as lubricants, and extra energy is generated by all the particles continually bouncing off each other.

In 1903, the east face of Turtle Mountain in Alberta collapsed and disgorged itself into the Crowsnest Pass below, travelling 2 miles (3 km) down the valley. As it hit the opposite side of the pass, the seething landmass swirled to a height of 450 ft (140 m), completely destroying the southern end of the town of Frank. When it finally settled, over a square mile (2.6 km^2) of the valley floor was carpeted with rubble up to 65 ft (20 m) deep.

Once a landslide has started, it is

NORTH AFRICAN AVALANCHE *The weight and speed of a rock avalanche makes it a most deadly form of mass movement, as can be seen in the Rif Mountains of Morocco.*

DAM BUSTER

In 1960, the Vaiont Dam in Italy was completed and the reservoir nearly full. The engineers did not realise, however, that the reservoir water was seeping into joints in the rocks of the surrounding mountains. They did notice that the soil was 'creeping' down the flank of one of the mountains at a rate of 10-12 in (25-30 cm) per week, but as this gradually reduced to 3/8 in (1 cm) per week they stopped worrying.

Then, in the late summer and early autumn of 1963, heavy rain raised the reservoir to more than 60 ft (18 m) above its normal working level, and the creep started again . . . reaching 8-12 in (20-30 cm) per day. Grazing animals abandoned the area and never returned. The engineers were slower to respond; on October 8, they began to drain the reservoir. It was too late. The following night, a mighty 9450 million cu ft (265 million m^3) section of mountain plunged into the remaining water. It all happened in less than a minute. Great waves surged beyond the limits of the reservoir, flooding lands far upstream. Downstream, the people of Crassa heard a sound like rolling rocks that grew to a thundering crescendo. A blast of air struck the town, breaking windows and lifting roofs. Then the wave struck; 60 ft (18 m) high, it gushed down the Longarone valley, carrying away five villages and killing 2600 people. The death toll made it the worst dam-related disaster on record.

DEATH VALLEY *Rescue parties search the ruins of the villages engulfed by the water that burst from the Vaiont Reservoir. Only a few people, living on higher ground, survived.*

unstoppable. But something must happen to agitate the slope in the first place and to start it moving. Earthquakes often provide just such an impetus, though water is the most common cause.

In August 1969, for instance, when hurricane Camille struck central Virginia, 27 in (69 cm) of rain fell in the Appalachians in eight hours. The sodden soil caused an epidemic of mud slides throughout the area with a death toll of more than 150 people.

LIQUID SOIL *Mud slides wash away roads and bridges on the Philippine island of Leyte.*

Although mass wasting is a natural process, it is often triggered or hastened by human activity.

THE HUMAN FACTOR

Exactly this had happened at Elm, where the quarry undercut and weakened the mountain slopes above it. Another example occurred on the Philippine islands of Leyte and Negros where two decades of illegal logging in the rain forests had boosted the local economy but left the deforested mountain slopes fatally exposed. When typhoon Thelma struck in November 1991, the screaming winds and torrential rain

NATURAL DAM *In India, in 1978, a landslide from the valley to the left of the picture blocked the gorge, creating a huge lake behind it (right).*

CANAL CLEARING *The Panama Canal, which is essential to world trade, needs constant dredging. Landslides from the surrounding mountains continually fill it with silt.*

ravaged the landscape throughout the islands, and soil no longer bound by tree roots or protected by undergrowth turned to slides of liquid mud. Half of the typhoon's victims were in the town of Ormoc on Leyte, where mud slides and floods claimed 3000 lives.

Two years previously, the Thai Government had banned commercial logging after landslides down deforested slopes had left 430 dead and 70 000 homeless. Unless similar action is taken in the Panama Canal region, the implications for world trade will be serious. Deforestation is baring the mountains to unprecedented levels of erosion, and debris washed down from the slopes is silting up the canal lakes. Unless this process is halted, the canal will become unusable one day.

The presence of trees increases the stability of shallow soil on a hill slope by 60 per cent or more. Not only do tree roots bind the soil, but the blanket of fallen leaves and undergrowth beneath them prevent the soil from drying out and crumbling away.

UNDERMINING NATURAL FORCES

Lowland areas and valley floors are accessible, sheltered and easy to manage, the obvious places for humans to cultivate and settle. As they become crowded, populations

THE RISE AND FALL OF A CIVILISATION

Gradual mountain-building processes can have as dire consequences for human life as the sudden devastating eruption of a volcano or the more blatantly destructive forces of a landslide.

At its peak, the powerful Chimu Empire, with a population of about half a million, stretched 620 miles (1000 km) from what is now southern Ecuador to central Peru. It grew up from about AD 900 on a fertile coastal plain where today's Moche river reaches the Pacific Ocean. The land must often have shuddered with the reverberations from volcanic eruptions and earthquakes, for it was on the brink of the region with the greatest density of active volcanoes in the world. Great mountain-building forces were at work – and still are today, as the oceanic Nazca plate is dragged beneath the continental South American plate; simultaneously, the overlying South American continent is forced up.

When the Chimu first settled there, the broad plain was a little above sea level and irrigation was not a problem. But as the land was gradually forced upwards – by about 20 ft (6 m) over 500 years – the Moche river cut downwards, forming a ravine below the level of the fields; the water table dropped, and the fertile agricultural land turned into semi-desert. The Chimu created extensive canal systems to carry water from the upstream reaches of the Moche and distribute it via a network of irrigation ditches across the plain. Aqueducts were driven through whole hillsides from neighbouring valleys. But all these efforts became redundant as the rising gradients stopped the water flow. The Chimu civilisation declined, and by the 15th century had disappeared. Today, even with modern farming techniques, only about 35 per cent of the land once so successfully farmed by the Chimu is productive.

LOST CITY *The ancient Chimu capital of Chan Chan in the Moche valley, near modern Trujillo, is now nothing but an archaeological site. The effects of mountain-building have left the land uninhabitable.*

spread to potentially more hazardous sites such as the slopes of mountains.

A natural slope forms gradually over the millennia through a process of uplift and erosion, creating its own equilibrium and stability. Slicing into this to build a road or a quarry immediately upsets the balance, and natural land-building forces will automatically try to restore it.

In Los Angeles, which is prone to landslides, huge basins have been excavated beneath the most unstable slopes as a safety measure. Just one of these basins might be the size of a football stadium, and there are 120 of them undermining the foot of the San Gabriel Mountains. They are continually being cleared of loose material washed down from the mountains, as the natural order seeks to reassert itself.

The city of Rio de Janeiro is crammed along steep valleys and coastal strips between precipitous hills and the sea. Over the last 40 years, the population has burgeoned and spilled over into shanty towns which cling precariously to the unstable sides of hills so steep-sided they are likened to sugar loaves. The heaviest rainfall on record, between January and March 1966, triggered landslides that swept through shanty towns killing nearly 300 people. A similar story can be told of Hong Kong island, where the mountain interior rises almost sheer above narrow coastal strips of flattish land. Yet pressure of space is forcing the population to spread uphill. Landslides on over-populated slopes claimed over 80 lives in 1972.

SLIDING CITY *The growth of Rio de Janeiro's shanty towns up the surrounding slopes results in landslides (below) and flood damage (bottom).*

ALL-ROUND SUPPORT

Tunnels in mines and for roads and railways used to be cut with arched roofs – so that they would take the weight of the overlying rock – but with flat floors. However, many incidents of shattered rock bursting up, through the floor, and killing workmen has brought home to engineers the fact that subterranean pressure comes equally from all directions. Tunnels are now excavated in circular cross-section to give all-round support.

FIGHTING BACK

In the notoriously unstable – and heavily populated – city of San Francisco, a 1982 landslide which killed 25 and caused $66 million-worth of damage, spurred the US Geological Survey to develop a warning system. Physical and environmental statistics, such as rainfall intensity, storm duration,

PREVENTING A ROCK FALL *Wire netting is placed across the potentially unstable slope of a road cutting in Hawaii to hold back any falling rocks.*

Slope Protection *Concrete and masonry walls (top) can stabilise a bare rock face, while mats (above) can prevent soil from washing away.*

slope, rock and soil conditions, were logged into a computer for analysis. Although the system was not fully operational in 1986, eight out of the ten landslides that year were predicted accurately. Warnings were broadcast, with bulletins on television and radio. Public response was limited. Even so, despite damage totting up to $10 million, only one person died and the exercise was regarded as a success.

Engineers are formulating guidelines for construction projects in hilly areas. They realise that a road cutting excavated at a sheer angle is more unstable than a gently sloping one, and that shelves or steps cut into a rock face may serve only to collect rocks which subsequently crash to the road below. To overcome these problems, overhanging rock faces can be propped up by concrete buttresses and rock bolts, or by steel tendons driven into the rock face to reinforce a slope. Concrete can be used to fill cracks or to spray-coat an entire surface, or a wire mesh cover may be fixed to contain loosened fragments. But these are cosmetic extras, compared with the most important preventative measure of all: adequate drainage. Water can be diverted down trenches (filled with gravel so that they do not erode and collapse). Cut a drain across a slope, and it may divert sufficient water for the water table to drop,

A Constant Threat *The sheer weight of moving rocks – as in this photograph of a boulder slide in the Wichita Mountains (right) – makes any kind of landslide difficult to control.*

allowing the slope to dry out and stabilise. Walls, including those made from boulder-filled wire baskets known as gabions, may sometimes reinforce a cutting, but drainage pipes are still necessary to prevent a build-up of water behind.

Rock Bolts *Steel rods driven through a sandstone rock face can hold it together (below).*

LARGE LANDSLIDES IN HISTORY

Year	Location	Extent	Trigger	Effects
1903	Alberta, Canada	1060 million cu ft (30 million m^3)	Mining	Town of Frank destroyed
1911	Usoy, Pamir, Russia	Not known	Earthquake	54 dead
1920	Kansou, China	30 000 sq miles (78 000 km^2)	Earthquake	180 000 dead
1927	Gros Ventre, Wyoming, USA	1400 million cu ft (40 million m^3)	Rain, snow melt	Blocked valley and river; town of Kelly destroyed
1959	Madison Canyon, Montana, USA	710 million cu ft (20 million m^3)	Earthquake	26 dead
1962	Huascarán, Peru	350 million cu ft (10 million m^3)	Earthquake	4000 dead
1963	Vaiont, Italy	9450 million cu ft (265 million m^3)	Reservoir fill	Displaced water; destroyed five villages; 2117 dead
1970	Huascarán, Peru	3500 million cu ft (100 million m^3)	Earthquake	Buried Yungay; 18 000 dead
1974	Mayunmarca, Peru	35 000 million cu ft (1000 million m^3)	Earthquake	Rock slide and debris flow; 450 dead
1985	Armero, Columbia	2100 million cu ft (60 million m^3)	Rainfall	5000 slides and debris flows; 333 dead

MOVING ICE

When it forms the fluffy flakes of the winter landscape, snow may seem benign. Yet tons of it, compressed together, produces a hard, unforgiving, material that we call ice. When this is on the move, disaster can follow in its wake.

The winter of 1950-1 saw one of the heaviest snowfalls that Austria and Switzerland have experienced since records were kept – up to 43 in (109 cm) in some places, during one four-day period.

On the high slopes of the Alps, snow built up to dangerously unstable levels. Piles balanced precariously on narrow ledges and overhung precipitous cliffs, until they suddenly toppled and hurtled down the mountains. Avalanches blocked all road and rail routes, and at one point in January 1951 the canton of Graubünden was completely cut off. On the night of January 20, a wave of avalanches hit the town of Vals killing 19 people. Farther west, not far from the St Gotthard Pass, a series of avalanches brought a reign of terror to the town of Andermatt. The first big slide came at dawn on January 21, thundering through the military hospital; the next two blocked the main road and swept away the railway bridge. In the early afternoon, the biggest avalanche of all smashed into the heart of the town, engulfing and destroying buildings, and burying 11 people alive. For the next three weeks, the town was plagued by slide after slide, as a further 6 ft (almost 2 m) of snow fell.

Many other villages in Switzerland and Austria suffered similar fates. By the end of what became known as the 'Winter of Terror', avalanches had claimed 279 lives, with hundreds more injured. Vast tracts of forest and cultivated land had been ripped up or swept away, and hundreds of head of livestock perished.

Fresh snow, made up of tiny powdery ice crystals surrounded by air, is so light that it takes about 10 in (25 cm) of snow to produce as much water as 1 in (2.5 cm) of rain. But if you pick up a handful of snow and press it into a snowball, the crystals are compacted and fused together, squeezing the air spaces out. Increase the pressure even more and the snowball turns to a hard lump of ice. On a grand scale, a similar compacting process can take place when a mass of snow slides downhill and hits the valley floor, creating an altogether more dangerous phenomenon than the powder-light crystals of falling snow would suggest.

SEEKING SURVIVORS *Rescuers sift carefully through the ruins of Vals (right) and Andermatt (top) in Switzerland after the 1951 avalanche disaster.*

SNOW KILLER

Avalanches occur after frequent snowfall, in mountainous areas where there are slopes of 30° to 40° for the snow to slide off. Snow will not slide off a gentler slope – although slush will slither down an angle of 10° or less – and will not settle in sufficient quantities on a steeper incline. How wet the snow is, and the degree to which individual crystals compact, gives rise to different types of avalanche.

CLOUD OF DEATH *An ephemeral-looking puff of snow disguises the deadly nature of an avalanche in the Swiss Alps (right).*

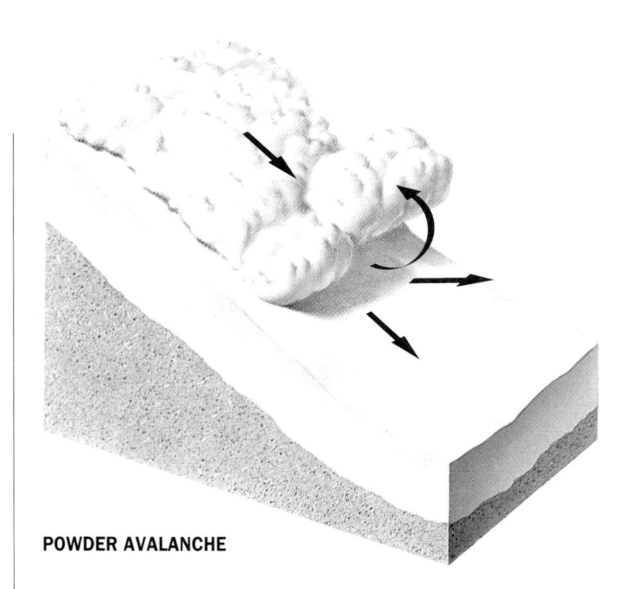

Powder Avalanche *Air trapped between the crystals of snowflakes (below left) gives them the lightweight layer that produces an avalanche (left).*

Freshly fallen, light, dry snow resembling very fine sugar crystals is the vital ingredient in what is known as a loose snow or powder avalanche – or in German as a *Staublawine* (dust avalanche). Usually, only the surface is affected, as individual particles roll downhill, dislodging others as they go like dry sand pouring down a slope. More and more grains are caught up over an ever-widening front and gather momentum to form a surging cloud of snow. If conditions are extremely dry and cold, the crystals do not fuse together at all. The snow is like powder and so light that it swirls several thousand feet into the air as the avalanche gathers power. The highly turbulent mass rolls down mountainsides at speeds of 110-180 mph (180-290 km/h) – and one in Switzerland even reached 225 mph (360 km/h). Powder avalanches like these often start where deep pockets of loose snow form in calm, windless conditions, or where the snow has been driven by howling winds to settle on the lee side of peaks.

The danger from a loose snow avalanche is not in its weight or volume, but in the shock waves that accompany it. An advance air blast of 100 lb per sq ft (0.5 tonnes per m^2) is followed by the maelstrom of particles within the avalanche itself, and together these are enough to disintegrate buildings. People caught in a blast of the turbulent air feel as if their lungs are bursting; others have been choked by dense flurries of powdery snow, as much as 1/2 mile (1 km) from the main path of the avalanche. In 1896, a powder avalanche blasted down a mountain into a Swiss valley with such momentum that it continued 2 miles (3.3 km) across the flat valley floor, and exploded several hundred feet up the opposite slope before settling.

Solid Masses of Moving Snow

A slab avalanche starts as a densely packed chunk of snow which can be longer than a football field and more than 30 ft (9 m) thick. It rides over a looser layer beneath, with tiny individual ice crystals acting like ball bearings to ease its way. The sheer weight and force of such slabs on the move makes them the most dangerous avalanches.

Unlike the arrow-shaped course of a loose snow avalanche, a slab avalanche has

Suffocating Snow *Dry light snow, usually freshly fallen, produces the suffocating sweep of a powder avalanche off Nuptse in the Himalayas.*

a broad front from the start, and leaves its birthmark behind it in the form of a horizontal or arc-shaped fracture line. Initially, the slab travels as a single carpet of solid ice which creases on its downward course – until it smashes into fragments with the violence and pressure of its passage.

SOLID SNOW *Snow that has lain for some time becomes compressed into a solid layer (below) which can turn into a slab avalanche (bottom).*

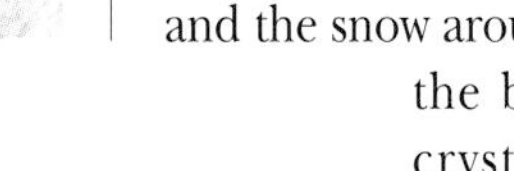

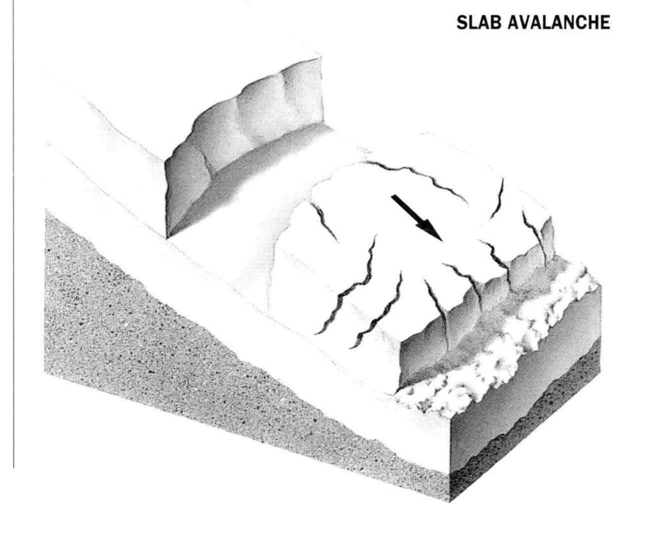

CRACKS OF DOOM *Arc-shaped fissures appear on a slope of compacted snow. The blocks of snow cut off by these become the slabs of a slab avalanche.*

Slab avalanches tend to occur on slopes exposed to prevailing winds which batter the surface layers of a snowfield and fuse the crystals into a compact mass. In a period when sunny days are followed by frosty nights, the crystals melt during the day and the resulting water refreezes at night, cementing the crystals into solid blocks of ice. Successive snowfalls under such conditions sometimes lead to layered blankets of snow which may slide off one another during an avalanche.

The spring thaw brings the danger of another kind of avalanche – a wet snow avalanche. Exposed rocks heat in the sun and the snow around them melts, weakening the bonds between individual crystals. Lumps of snow like loose snowballs detach themselves and increase in size as they bounce down the slopes, becoming heavier and more compact as they go, and eventually solidifying when they come to a halt. The massive weight of an avalanche like this wrenches boulders and mature trees from the mountain slopes, and scours the surface beneath of its vegetation and soil. In 1885 a wet snow avalanche in Italy snowballed into 2.5 million tons of snow and rocky debris; another in the Vinadi region of Switzerland in 1962 left larch and pine trees snapped off at their trunks.

A variation on this kind of avalanche occurs when a rapid thaw is combined with heavy rainfall. Water seeps into the snow and saturates the bottom layers, to create an aptly named slush avalanche or *Grundlawine* (ground avalanche).

Long after the snows have gone, the scars an avalanche leaves in its path remain. These may include slopes stripped of all permanent vegetation or, less drastically, strips of stunted shrubs and small plants lying between spurs of sturdy conifers. Untidy piles of rock lie at the feet of the slopes, their jumble of shapes and sizes a sure sign that they were dropped suddenly as the avalanche came to a halt. These rocky screes, as the piles are called, fan out from the base of the slopes, reflecting an avalanche's course as it slowed and spread before settling.

Another sign of an avalanche is a hollow or small lake, at the foot of a steep-flanked glaciated valley, which may have been gouged out by successive avalanches. High

AVALANCHES THAT CHANGED THE COURSE OF HISTORY

The Carthaginian general Hannibal had an audacious plan to conquer Italy, starting with a surprise attack from the north via the seemingly impenetrable barrier of the Alps. His plan failed, not through battle losses but from avalanches. The first snows had already fallen in early October 218 BC, when Hannibal led his 38 000 soldiers, 8000 horses and 37 war elephants into the mountains. By the time they reached Italy, nearly half of the soldiers, 2000 horses and many of the elephants had perished in the snows. The remainder of the army was exhausted, and after a few early successes it was defeated and Hannibal committed suicide.

On December 13, 1916, at the height of the First World War, a series of avalanches in the Italian Dolomites enveloped the opposing armies of Austria and Italy, killing as many as 10 000 soldiers in the space of a day. It has been estimated that more than 40 000 perished in Alpine avalanches during the First World War. Many of these slides were strategically induced – by blasting snowbanks and cornices to cascade over enemy positions.

ALPINE JOURNEY *In this 16th-century Italian fresco, Hannibal – one of the great military leaders of antiquity – is shown crossing the Alps, a journey beset by avalanches.*

on the mountain slopes, there may be hollows and scrape marks in the rock, overturned trees, or pits from which trees or boulders have been torn.

When skiers hear the roar and see the rolling clouds of an avalanche advancing towards them, they generally shed surplus equipment immediately, point their skis downhill and race for their lives. They know that their chances of survival if caught are fairly slim.

PATH OF DESTRUCTION *The full effect of damage to trees caused by an avalanche becomes visible in spring after the snow has melted.*

CHANCES OF SURVIVAL

A fifth of those who get caught die instantly during the initial impact of hurtling ice, snow and rocks. If they survive that, they may be able to ride the avalanche, using swimming motions to keep on the surface. If buried, they have to contend with cold and immobility. The victim's only chance lies in clearing a breathing space around his or her face and hoping that rescue comes soon.

According to the statistics, rescue teams find only about 5 per cent of avalanche casualties alive. The main difficulty is in tracking the victims down. Rescue teams move in a thin line in tight formation (a human body is narrow and easy to miss), probing the snow for bodies with long poles called sounding rods; many ordinary householders in an avalanche area have one at home. The poles are up to 15 ft (4.6 m) long, and sometimes consist of lightweight sections, like tent poles, that make them easier to carry and assemble on the spot.

The rescuers also look out for pinpoints of bright red safety cord that the wise skier carries with him and pays out as soon as an avalanche is spotted. Teams may carry magnetometers to detect minute variations in magnetic fields and to register the presence of metallic equipment such as skis, or infrared detectors to locate body warmth. Specially trained dogs – traditionally St Bernards, but now often German Shepherds – can search an area in half an hour that would otherwise take 20 people four hours to cover as thoroughly.

Avoiding avalanche-prone slopes is an obvious safety measure. Experienced mountaineers avoid cornices – peaked caps of snow that overhang cliff faces or that teeter on the leeward slopes, where strong winds have driven flurries and whipped them into

Rescue Operations
Avalanche rescues feature teams of people standing shoulder to shoulder, probing for survivors with long poles (top), and sniffer dogs (above).

knife-edged sculptures. It may take no more than a shout to dislodge them.

Faced with a suspect slope, skiers cross at high levels – above the likely starting point of an avalanche – and make the journey one at a time, at least 50 yd (45 m) apart. They free their hands from the loops on their ski sticks, and undo their safety straps to reduce the risk of entanglement.

At winter sports resorts, weather predictions, avalanche alerts and regular inspections of the slopes reduce the numbers of people risking danger areas. The Swiss Federal Snow and Avalanche Research Institute runs an Avalanche Warning System, based on meteorological observations throughout the country. Japan's Snow Research Station has devised a scale, similar to the Richter scale for earthquakes, to describe the effects of individual avalanches.

Moving to an Avalanche Zone

Home-builders in avalanche-prone areas have developed their own safety measures. In Switzerland they consult a Risk Map which charts avalanche probability in each area, based on past statistics. Many Swiss residents build their houses in the rough shape of a ship, with the prow end pointing uphill. Then, if an avalanche hits, instead of crashing against a flat wall at full impact, it will part and slide to either side of the 'prow'. Swiss building regulations demand that all new buildings are strong enough to withstand an impact of 3 tons per sq yd (approximately 3 tonnes per m²). Originally, this applied only to roofs and walls facing the course of an avalanche, but it is now known that avalanches have a drag effect which can pull a building along as it passes, so that strong foundations are equally important.

As far back as 1716, the villagers of Pequerel in the Italian Alps took anti-avalanche measures, erecting a giant

AVALANCHE RUSH

The greatest concentration of North American avalanche casualties was during the Gold Rush in the Rocky Mountains between 1860 and 1900. Thousands of prospectors poured into the Rockies in their quest for gold, unprepared for the mountain environment and its dangers, and ignorant of the impact their numbers and gold-digging would have on it. Many avalanches were triggered by their activities – for instance, the series of avalanches that engulfed prospector encampments around Alta, Montana in 1874, claiming 60 lives.

Warning Sign *Wind and thaw have removed part of the snow from the peak of the Jungfrau in Switzerland, leaving the remainder as unstable overhangs and cornices.*

Danger Areas *Avalanches on Mount McKinley in Alaska (above) and Lhotse in the Himalayas (left) demonstrate why cornices and steep slopes should be avoided in mountainous areas during the avalanche season.*

wedge-shaped rock wall above the village. Today, after many reinforcements, it stands 6 ft (1.8 m) thick, 16 ft (5 m) high and protects a 600 ft (180 m) stretch. Another tale of avoidance measures involves the single worst avalanche disaster in American history. Three trains sat idly in the station at Wellington in the Cascade Mountains in Washington State, USA, on March 1, 1910. Late season snowfalls had blocked the nearby Stevens Pass, and passengers on their way to or from the coast at Tacoma were getting cold and irritable, wondering how much longer they would be delayed. Suddenly, an immense bulk of snow detached

DAMAGE LIMITATION *In avalanche areas, the roofs of buildings are constructed flush with the slope of the hill (top), roads are protected by tubes (middle), and snowfields are stabilised by fences (above).*

itself from the mountain above and crashed down, picking up all three trains and passengers, and the entire station, and depositing them 150 ft (45 m) over the rim of a canyon. A hundred and eighteen people were killed. The railway track was relaid through tunnels, for avalanches are unstoppable and avoidance and protection measures are the most effective way to counter them. A less costly alternative is a surface tunnel like the roof of a shed, which continues the line of the mountain slope over the road or railway cutting, so that the avalanche passes over and continues down the other side.

The 1950-1 winter of terror prompted further serious study and the development of preventative measures, based on the three stages of an avalanche's course: the starting zone, the track and the final 'runout' zone.

Avalanche Prevention

In the starting zone, fence-like wooden baffles or aluminium strips mounted on tripods are designed to disrupt windflow. This has the effect of diverting the falling snow so that it forms irregular patches rather than dangerous accumulations. Snow rakes – fences of wood, steel or aluminium poles – prevent large areas of snow forming into unstable slabs.

In the track zone, sturdy walls and wedges are placed to deflect the course of an avalanche, and shed-roof tunnels are constructed for the avalanche to pass over harmlessly. Walls are constructed to deflect avalanches around built-up areas in the runout zone, and mounds of soil or masonry piled along valley floors help slow the progress of a spreading avalanche.

Experiments with antifreeze are still in their infancy, but chemicals can be sprayed upon a snowfield to break down the crystals and melt the snow, or onto a bare surface to prevent the snow from settling.

The most effective barrier of all, however, is a natural one: a forest of big, mature trees growing tightly together. Unfortunately, most of the original Alpine forests were cleared for timber centuries ago, and establishing new plantations in the areas where they are most needed is difficult, for these are also the areas where saplings are most likely to die from extreme cold or to be uprooted by an avalanche. Selected areas of the remaining Alpine forests are known as the *Bannwald* (forbidden forest), and entry is banned in a bid to preserve them.

Explosives help to prevent avalanches, too. Instead of waiting for vast unstable amounts of snow to build up, pressure can be released by deliberately setting off a series of small, harmless avalanches, using hand-thrown explosives or projectile mortar bombs. Explosives are sometimes planted in known avalanche zones, well before the winter season sets in, and then detonated at various stages throughout the winter to prevent dangerous build-ups. The Swiss originally developed this technique, known as explosive avalanche induction, but the Americans have diversified the equipment used in the Rocky Mountains, deploying 3 in (75 mm) howitzers and 3 in or 4 in (105 mm) recoilless rifles,

CONTROL MEASURES *Specially designed artillery is sometimes used to trigger avalanches in a controlled manner.*

A FENCE OF FOREST *Trees stabilise snowfields and form natural barriers to avalanche movement.*

PRESERVED BY AN AVALANCHE

In 1991, the fully clothed body of a shepherd who lived more than 50 centuries ago was discovered perfectly preserved in a melting glacier. Dust carried from the Sahara desert by warm, southerly winds had settled on the glacier surface. It absorbed the warmth of the Sun, hastening the melting process.

Travellers in the Ostaler Alps, between Austria and Italy, were crossing the Similaun Glacier when they saw the shape of a human body in the ice beneath their feet. It turned out to have lain there for 5300 years and was the oldest complete human body ever found.

Experts surmised that the man had been caught in a blizzard, had sheltered in an icy hollow and had frozen to death. His body must have lain exposed to the winds for long enough to become dehydrated and to ensure its remarkable state of present-day preservation, but was then probably buried by an avalanche and successive layers of snow and ice until its discovery.

AN EARLY VICTIM *The body of a man buried in snow and ice for more than five millennia thawed out of a glacier in 1991.*

or custom-built compressed-air cannon. The main problem with avalanche induction is that even the smallest avalanches can get out of control.

CREEPING ICE AND FLOATING ICE

Glaciers have neither the speed nor sudden drama of an avalanche, but they do have the power to change the face of the landscape. When snow collects in depressions high in the mountains and does not melt from one winter to the next, it builds up and the weight of successive layers compresses it into ice. Eventually, the accumulation of packed ice spills sluggishly over the lip of a valley head. It is frozen solid, weighty and as

SLOW BUILD-UP *Glaciers creep slowly down valleys (left) or across flat plains and water surfaces (above).*

dense and viscous as a lump of putty.

A glacier's slow, tongue-like advance belies its erosive force, as it gouges out the valley floor and planes back the walls, transforming a sharp, V-shaped valley profile into a broad-bottomed, steep-flanked 'U'. Its upper surface is creased and cracked from the stresses of negotiating the irregularities of its mountain course, and it is littered with lines of debris from rock falls and avalanches.

The longest valley glacier in North America is the Hubbard, which runs 93 miles (150 km) from the interior of Alaska to the Bay of Disenchantment on the Gulf of Alaska. Its normal pace is a leisurely few inches a day. But in early 1986, it suddenly, and so far inexplicably, speeded up to about 150 ft (45 m) per day for a month or so. Normally, the glacier melts at its sea end as fast as it grows at the mountain end. Now it was not melting fast enough, and the snout advanced the length of the Bay of Disenchantment before settling like a whale at the entrance of the tributary Russel Fiord and blocking it. The repercussions on the local economy which depended on Russel Fiord's fishing industry were dire, and there were wider environmental implications as well. The water trapped behind the glacier in Russel Fiord was constantly fed by freshwater mountain streams and its level was rising by nearly 1 ft (30 cm) a day. What should have been a salt-water inlet was becoming a freshwater lake, completely upsetting the area's ecological balance and threatening the survival of some 40 porpoises, harbour seals and sea lions that had been trapped. In the event, the problem resolved itself; by autumn, the end of the glacier had melted.

AVALANCHES THAT HIT THE HEADLINES

Year	Location	Comments	Deaths
1902	*Telluride, Colorado*	*3 avalanches hit in one day*	*19*
1910	*Wellington, Washington*	*Trains and station swept away*	*118*
1914-18	*Italian Dolomites*	*Some were deliberately induced by combat troops*	*40 000*
1926	*Bingham Canyon, Utah*	*Mining town demolished*	*40*
1950-1	*Switzerland and Austria*	*'Winter of Terror'*	*279*
1954	*Vorarlberg, Austria*	*Warning system failed*	*125*
1962	*Nevado de Huscarán, Peru*	*3 tons of snow and ice fell into the valley*	*4000*
1965	*Mattmark, Switzerland*	*Ice fall from glacier*	*88*
1970	*Yungay, Peru*	*Earthquake triggered*	*2000*
1978	*Col des Mosses, Switzerland*	*Ski lift overwhelmed*	*60*
1979	*Lahaul Valley, India*	*Snow from Himalayan foothills buried the valley*	*200*
1997	*Mazar-e Sharif, Afghanistan*	*Swept away bus passengers*	*100*

FLOATING MENACE *The iceberg that poses such a threat to shipping in the north Atlantic began its life as a snowfall.*

When glaciers reach the sea, they break up, and disconnected chunks of ice float away as icebergs. Every year, some 12 000 icebergs are 'calved' from the Greenland ice cap into Baffin Bay. Numbers vary from year to year: in 1958 only one Baffin Bay iceberg reached the Atlantic shipping lanes; the following year, there were 693. Their progress is logged by the International Ice Patrol, for icebergs are a shipping hazard, especially if they reach the Newfoundland coast with its treacherous fogs – the 'unsinkable' *Titanic* being their most famous victim, in 1912.

At the start of its journey, an iceberg may weigh 1.5 million tons, rise 260 ft (80 m) above sea level and plunge more than 1200 ft (360 m) beneath the surface. By the time it reaches the Atlantic, it will have lost weight and be sculpted by melting and erosion into fantastic shapes, often with lethal ridges concealed below the surface.

3 The Restless Air

Sinister Swirl *A camera in a space shuttle captures the side of an Indian Ocean cyclone.*

The atmosphere is in constant turmoil. On a global scale, it acts as a giant heat exchange, in which the hot air of the tropics rises and moves towards the poles, while cooler, heavier air from the poles rushes in to take its place. Averaged out over the years, these movements form the conditions that we call climates. Day-to-day variations in these climatic patterns produce what is known as weather. These variations can be extreme, including hurricanes that cut a great curving swathe across the sea — and sometimes the land as well — and deadly tornadoes, those narrow funnels of extreme low pressure and extraordinary destructiveness.

Vicious Spiral *A tornado advances across the prairies.*

HURRICANE HORROR

Hurricane-force winds may be encountered all over the globe, but it is only above the warm seas of the tropics that a ripple of instability in the air can become a genuine hurricane, the deepest of all low-pressure weather systems.

Galveston, Texas, is a major seaport on the Gulf of Mexico. By the start of the 20th century, the thriving cotton trade of the hinterland and the port's growing popularity as a holiday resort had made it one of the most prosperous cities in the United States. The town is built on an island, one of the many sandbars created by currents sweeping clockwise around the Gulf. Galveston's highest point is only 6 ft (less than 2 m) above sea level.

A tropical storm that developed in the Atlantic Ocean in the early part of September 1900 seemed too distant to be of any relevance to the people of Galveston. Even when warnings were issued after the storm had hit some Caribbean islands 600 miles (960 km) to the east, Galveston was still basking in warm, late summer sunshine, a few wispy clouds high in the sky. Slowly, however, the clouds thickened and lowered, and a fresh wind sprang from the east.

Early on the morning of September 8, it started to rain and sizable waves were beating upon the shore. The whole surface of the sea seemed higher than usual, and as the wind shifted to the north and strengthened, the waters began to flood into the low-lying streets. People abandoned their houses and tried to leave the island, only to find that the single bridge to the mainland and safety had been covered by rising water. By noon the wind was a fierce 40 mph (64 km/h) and still increasing. The streets were under 4 ft (1.2 m) of water and it was still raining. By mid afternoon, half the sandbar island was under water and waves were pounding against the buildings, weakening their sandy foundations. Wooden buildings collapsed and were swept away with their occupants. Some fled their homes and made for the city's stone-built Sacred Heart Church. Electricity generating stations were destroyed and a complete blackout added to the howling chaos as night closed in. Building after building disintegrated in the face of winds estimated at 110 mph (177 km/h) and waves 5 ft (1.5 m) high.

Suddenly everything fell quiet and a ghostly calm settled over the ravaged city. People peered from their shelters to see stars in a cloudless sky, and thought that it was all over. The lull lasted a few minutes; then, as suddenly as it came, it ended as the driving rain and screaming winds returned. This time they came from the opposite direction, the south. More buildings were wrenched away and carried out to sea, so that by morning, when the hurricane's fury had abated, not one building remained intact in at least half of the town. All that was left of the Sacred Heart Church and the 400 who had taken refuge there were two steeples and an empty shell of broken walls. The subsiding waters left thousands of corpses – people and animals – scattered along the beach. The most deadly hurricane in the history of North America had taken 6000 lives in Galveston – about 15 per cent of the population – and a further 6000 in the surrounding mainland.

TERROR IN TEXAS *The seaport of Galveston was the site in 1900 of one of North America's greatest hurricane disasters.*

SECOND STRIKE *Galveston (right) was once again – in 1983 – the victim of a violent cyclone, Hurricane Alicia.*

SETTING THE SYSTEM IN MOTION

Hurricanes and tornadoes are extreme manifestations of the basic principles that guide the world's wind and weather systems. The atmosphere is constantly in motion, flowing like ocean currents above the surface of the planet. Areas of the Earth's surface are heated by the Sun and transfer heat to the air above them. As air warms, it expands, becomes lighter and rises, to be replaced by cooler air moving in beneath. As air cools, it becomes denser and so sinks and flows outwards. A constant exchange or cycle of warm and cold air is set up, and the flow generated is the wind.

On a global scale, hot air blows towards the poles from the tropics and cold air blows towards the Equator from the tropics.This is because the hot air around the Equator rises and is replaced by cooler air from the tropics that is sucked in to replace it. If the Earth did not rotate, these flows of air would be more or less north-south. Instead, winds heading for the equator are deflected westwards, and those blowing away from the Equator are deflected eastwards by the Coriolis effect (named after the French mathematician Gustave-Gaspard Coriolis who first described it in the 19th century). This is the effect of the Earth's rotation on the movement of the atmosphere. It means that winds travel in a curve rather than in a straight line and, in the localised context of hurricanes and tornadoes, that winds are sucked inwards on a curve rather than flowing straight.

Imagine standing at the centre of a slow-spinning turntable and throwing a ball to someone at the outer edge. By the time the ball reaches the edge, the person to whom you were throwing it will have been left behind by the spin of the turntable. The ball will appear to have followed a curving course away from the receiver. Under the influence of the Earth's rotation, winds – like the ball – flow in a curving path.

The Earth spins eastwards on its axis once every 24 hours. Around the poles, as around the centre of the turntable, the land does not move much to make a complete turn in the space of a day. But farther away from the poles, as the planet broadens, more land has to shift in the same amount of time, with the result that the spin rate increases, until a point on the Equator will be rotating at more than 1000 mph (1600 km/h). A warm air particle moving away from the Equator will, therefore, have an eastwards momentum to start with, relative to the more slowly moving land beneath it. It will continue to curve eastwards from the Equator. Conversely, colder air particles moving from the poles towards the Equator will travel over a surface that is moving progressively faster and will, therefore, be deflected to the west.

TURNING EARTH *Under the influence of the Earth's rotation, winds curve to the right in the Northern Hemisphere and to the left in the Southern Hemisphere: a phenomenon known as the Coriolis effect.*

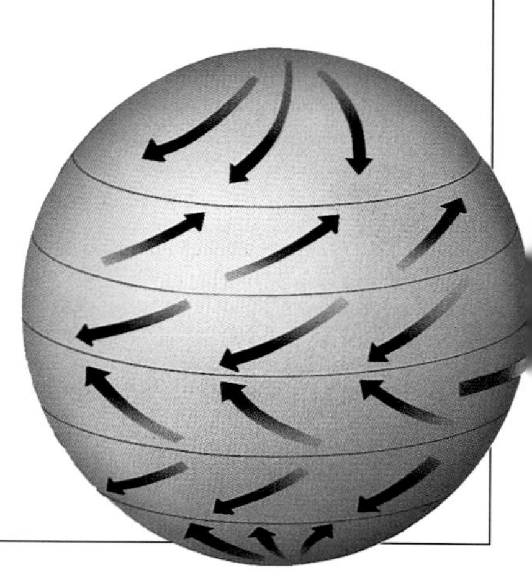

The victims were buried at sea to avoid the spread of contagious diseases. The mainland's cotton crop was destroyed, sending severe economic repercussions throughout the textile industries of Europe. Galveston was rebuilt with a defensive sea wall, which proved its worth when a hurricane hit the city again in 1915 with only a dozen or so victims.

Hurricanes are spawned in a narrow band of ocean between latitudes 5° and 20° north or south of the Equator, where the warm tropical seas have a surface temperature of at least 26°C (79°F) and the trade winds carry air away from the Equator.

The Atlantic hurricane (from the Carib Indian word *huracan*, 'big wind'), the Pacific typhoon (from the Chinese *tai feng*, 'big wind') and the tropical cyclone of the Indian Ocean are all the same thing. The British sea captain Henry Piddington first coined the word cyclone in the 1840s to reflect the hurricane's circular, whirling character, and it has since become the accepted meteorological term. (An anticyclone is an area of high atmospheric pressure, in which winds spiral outwards rather than inwards.)

HOT CLIMATES – BIG WINDS

The whirling forces capable of destroying whole towns may begin as a mere ripple of unstable air over warm tropical seas. In the extreme heat of the tropics, the air expands, becomes lighter and rises, leaving a patch of intense low pressure below. Cooler, denser air is sucked in from the surrounding atmosphere, not in a straight flow, but

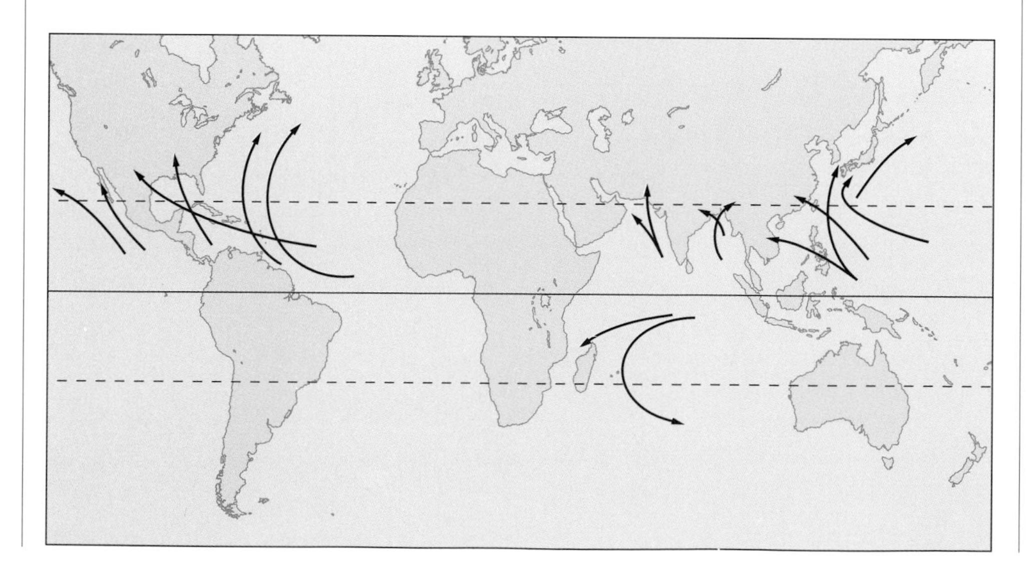

MAKING TRACKS *Hurricanes originate in latitudes between 5° and 20° north or south of the Equator, and then move slowly westwards.*

HURRICANE STATISTICS

Diagrams of hurricanes are usually exaggerated in height to illustrate the mechanics in detail. In reality, a hurricane is squat, its height (which may be greater than Mount Everest) about 1/24 of its width. The average diameter of a fully mature system is 260 miles (420 km), although some can be as much as 400 miles (644 km) – and the eye can be up to 30 miles (48 km) across.

in a spiralling action because of the Earth's rotation. A tropical storm develops, consisting of a fast-whirling current that sucks air into its centre at sea level and draws it upwards in a spiral around a calm core.

Weather experts believe that a contrast in temperature between an evaporating and cooling ocean surface and the warm air above it creates the unstable conditions in which this process can begin. When rain from moisture-carrying trade winds falls over warm waters, latent heat – heat released by a substance as it changes its state – is

FROM THE SAFETY OF ORBIT *A satellite captures the full extent of a hurricane's swirl.*

BENEATH A HAZE OF CLOUD *North America awaits the onslaught of a hurricane spiralling in the Gulf of Mexico.*

ANATOMY OF A HURRICANE

Wispy cirrus clouds of ice particles

Spent winds spiral outwards

Cool air descends in the eye

Thick cumulus clouds of water droplets

Rising hot air

Extremely strong winds spiral into the low-pressure centre

A bulge appears as the ocean surface is drawn up by the low pressure

Scale diagram of a hurricane's horizontal extent

HURRICANE ANATOMY *The scale of the vertical axis has been exaggerated to emphasise the air movements in a hurricane.*

CLUES FROM A WEATHER MAP

As the contour lines of a map give an impression of the shape of the land, so isobars – lines which link areas of equal air pressure – reveal the highs and lows of the weather on a meteorological chart. The closer together the isobars are – like the contour lines of a steep slope – and the more there are of them, the greater the gulf between the depression and the normal atmospheric pressure around, which gives rise to very intense winds. Continuing this comparison with relief maps, meteorologists refer to 'pressure gradients'.

released by the condensation. This gives a boost to the surrounding air temperature; the surface of the sea, by contrast, cools as water evaporates from it.

The moist, warm air rises until the air pressure drops with height and thick banks of cloud form. At the point where the moisture can no longer be held as vapour, it condenses and falls as torrential rain. Increasingly strong winds are sucked into the low pressure at the foot of the pillar of rising air, and once the winds reach 74 mph (120 km/h), the storm is classified as a hurricane. The vortex of thick, rising cloud can tower to heights of 6 miles (10 km) above sea level, flickering constantly with internal lightning. The winds inside can spin at a sustained 74-186 mph (120-300 km/h), though the system as a whole travels at 15-50 mph (24-80 km/h).

As the vortex reaches a height of about 30 000 ft (9150 m), the hurricane clouds are flung outwards because the density of the warm air equals that of the atmosphere at this height, and so the clouds lose their bouyancy. This final, outward spiral has been captured in several dramatic satellite photographs.

In the centre, or eye, of a hurricane is an area of eerie stillness with clear skies and high temperatures; which may measure as little as 1-2 miles (1.6-3 km) in diameter or as much as 30 miles (50 km). The brief interlude of calm in the Galveston hurricane was the passage of the eye of the storm. At the edge of the eye are the fiercest winds of all, raging at up to 220 mph (354 km/h), with the result that the transition from calm to chaos as the eye passes over can be extremely sudden.

It takes several days for a hurricane to travel from its ocean birthplace to land where it can do any serious damage – the average life of a tropical cyclone is 7-14 days. Although this gives plenty of time for warnings to be issued, the exact track of a hurricane is always erratic and hard to predict. Hurricane Gilbert roared into the Caribbean from the Atlantic in September 1988. After striking Yucatan on the southern fringe of the Gulf of Mexico, it swung north towards Texas. But forecasts differed; some believed it would head towards northern Mexico. The people of Galveston, Texas, took no chances; they evacuated the city. The hurricane went to Mexico.

A Trail of Destruction

The winds of a hurricane bend palm trees until their fronds stream out like windsocks. Boats are lifted from harbours, flung over sea walls and smashed to splinters far inland. Sturdy houses disintegrate as their roofs are wrenched off and their foundations scoured away by giant, wind-whipped waves. Sand and rubble are deposited over roads; power and communications lines are ripped from their poles. If the turbulent thunderclouds of a hurricane generate a tornado – which 25 per cent of them do – the chaos increases. On the other hand, hurricane damage is confined to islands and coastal areas. The

Stormy Weather *Crashing waves and bending trees are typical of hurricane strikes, as in cyclone David (above) and hurricane Allen (right).*

HURRICANE DAMAGE *In 1989, Hurricane Hugo caused physical damage, including overturned planes in airports, that amounted to $1 billion in Puerto Rico.*

storm's fury abates quickly over land, mainly because it runs out of the energy and moisture that it gets from warm water. Within tens of miles from the coast, the whole wind system collapses upon itself and dies.

The force of a hurricane raises massive areas of sea, with water surges bringing wild breakers and severe flooding. The difference in atmospheric pressure between the extreme low at the base of the hurricane and the higher pressure of the surrounding air causes the sea surface within the hurricane system to be drawn upwards like iron filings to a giant magnet. The water level within the system can be 15 ft (4.6 m) higher than that of calmer seas around.

The Great Atlantic Hurricane, which swept up the eastern coast of the USA from Puerto Rico to New England in 1944, left evidence of dramatic surges. Tide meters along the coast recorded three stages. The first was marked by a shift in normal tide patterns, caused by atmospheric movement far in advance of the hurricane. In some places, the water level rose abnormally; in others, it fell.

The second stage was the actual hurricane surge, a sharp rise in sea level that one eyewitness mistook for a thick bank of fog rolling in from the ocean. A hurricane surge may last only a few hours – but its crashing waves wreak severe damage. After the hurricane come the 'resurgences', a series of smaller waves like the wake of a ship. Although resurgences are a shadow of the main surge, their effect can be amplified by local effects such as high tide, and they often catch survivors unawares.

MEASURING A HURRICANE

The intensity of a hurricane is measured on the Saffir-Simpson scale. There are five levels of intensity:

- Level 1: wind speed 74-95 mph (120-153 km/h), weak
- Level 2: wind speed 96-110 mph (153-177 km/h), moderate
- Level 3: wind speed 111-130 mph (177-209 km/h), strong
- Level 4: wind speed 131-155 mph (209-250 km/h), very strong
- Level 5: wind speed over 155 mph (250 km/h), devastating

ALL-TIME LOW The lowest atmospheric pressure ever recorded was 870 millibars in the heart of Typhoon Tip in the Pacific Ocean in 1979. This is about 85 per cent of normal atmospheric pressure.

Most of the damage inflicted by the hurricanes – or tropical cyclones – of the Indian Ocean is from water surges, because much of the land surrounding the ocean is low-lying. The nation of Bangladesh, for instance, is built on mud flats, where vast Asian rivers such as the Ganges and the Brahmaputra converge in the world's biggest delta before entering the Bay of Bengal. Indeed, only a quarter of Bangladesh lies more than 10 ft (3 m) above sea level, yet it supports 117 million people and is the eighth most densely populated country in the world. People set up home on the fertile but temporary banks of silt, called *chars*, which are deposited by the sluggish rivers, and cultivate them intensely with rice and jute.

THE HUMAN COST *Hurricane Hugo left 18 people dead and more than 4700 injured.*

Resort Ruins *Hurricane Andrew wreaked havoc amongst holiday homes and caravans in Florida in 1992.*

The tropical cyclone that gathered in the Bay of Bengal in May 1985 was detected well in advance, but few Bangladeshis could afford radios on which to hear the warnings. Others were reluctant to leave their homes to squatters and looters, and besides, there was no place to move to in such a crowded land, and certainly no ground above flood level.

As the storm gained in force, the surface of the ocean was gathered up by the sinking atmospheric pressure and whipped by hurricane winds into giant storm waves 15-50 ft (4.6-15 m) above normal sea level. This monstrous sea poured into the Bay of Bengal and surged over the lands of the delta.

Beached Boat *Coastal areas suffered the most damage from Hurricane Andrew.*

Torrential rain brought flood water from inland areas, swelled by water and mud streaming from severely deforested hill slopes. The death toll was somewhere between 15 000 and 100 000 (statistics of both living and dead are difficult to gather in Bangladesh), and an estimated 250 000 were left homeless. Entire settlements had been washed into the sea; crops, fishing boats and 30 000 cattle had vanished. Survivors clung to bamboo rafts and thatched roofs, fending off crocodiles and sharks.

For Bangladesh, the disasters wrought by typhoons are an ongoing saga. Since records began, there have been 60 devastating cyclones, with an estimated total death toll to 1985 of 1.6 million. The worst individual disaster was November 13, 1970, when between 500 000 and 1 million people died. In May 1985, up to 100 000 perished, and in April 1991, 138 000.

The typhoons of the north Pacific affect Japan, China and, particularly, the Philippines – where 20 storms a year are common. In Hong Kong's 1937 typhoon 11 000 people died. When Typhoon Vera hit Nagoya in Japan in 1959, one storm surge rising in the shallow Ise Bay claimed 5000 lives. A 1991 *baguio* – the Philippines name for typhoon – was directly responsible for 3000 deaths. A further 3000 people drowned in liquid mud, which the typhoon rain sent slipping down mountain slopes that had been left bare by illegal logging.

Hurricanes of the Southern hemisphere occur in patches around Madagascar in the Indian Ocean, and along a broad swath stretching from the north-east coast of Australia and across the South Pacific, encompassing the island chains of Melanesia, Micronesia and Polynesia. In the South Pacific, Papeete, the capital of Tahiti, was pounded by 65 ft (20 m) waves in February 1906. The same hurricane hit the Society and Cook Islands with a resulting death toll of around 10 000.

The islands of Micronesia, Tuvalu and Kiribati are particularly vulnerable to even moderate sea surges, because they are little more than coral reefs breaking the surface of the ocean with high points only a few feet above sea level. Fortunately, they are in the relatively safe territory where hurricanes breed – by the time the storms have

WHAT'S IN A NAME?

The practice of naming tropical cyclones after people was started in the 1890s by an Australian meteorologist, Clement Wragge. At first, he used letters from the Greek alphabet and characters from Classical mythologies, but later he chose the names of local politicians. When they threatened legal action, Wragge used their wives' names instead. Naming was adopted by the World Meteorological Organisation in 1952. Only female names were used initially, but for the sake of sexual equality, male ones have also been used since the 1970s – in alphabetical sequence through the year.

LOW-LYING LAND *Bangladesh lies right on the track of the Indian Ocean cyclones.*

reached their full powers of destruction, they have usually travelled much farther west to the East Indies and the north-east coastline of Australia. The tropical cyclone that devastated the town of Darwin in Northern Territory in 1974 originated in this way far out in the ocean.

Other, more typical cyclones are more local affairs. They originate in the Timor Sea between northern Australia and Indonesia and move south, causing much damage to the coastal communities of northern Australia. The cyclone also has a positive effect – it brings rain to the dry

WIND-DRIVEN WATERS *Flooding is the worst result of a cyclone hitting Bangladesh.*

MAJOR HURRICANES, TYPHOONS AND CYCLONES

Date	Area of Hurricane	Death toll (est.)
1893	*Louisiana, USA*	*2000*
1900	*Galveston, Texas, USA*	*10 000-12 000*
1906	*Society and Cook Islands*	*10 000*
1928	*Okeechobee, Florida, USA*	*1836*
1938	*New England, USA*	*600*
1957	*Louisiana, USA*	*390*
1959	*Japan*	*5000*
1960 (Oct)	*Bangladesh (then East Pakistan)*	*6000*
1963	*Bangladesh (then East Pakistan)*	*2000*
1963	*Cuba and Haiti*	*7000*
1964 (May)	*Bangladesh (then East Pakistan)*	*35 000*
1964 (Dec)	*Bangladesh (then East Pakistan)*	*15 000*
1965 (June)	*India and Bangladesh (then East Pakistan)*	*30 000*
1970	*Bangladesh and India*	*1 million*
1974	*Honduras*	*2000*
1977	*South-eastern India*	*100 000*
1979	*Caribbean and USA*	*1100*
1985	*Bangladesh*	*15 000-100 000*
1991	*Bangladesh*	*138 000*
1991	*Philippines*	*6000*
1992	*South-eastern USA*	*58*

interior of Australia. Japan, too, depends on typhoons for a quarter of its rainfall.

The study of hurricanes is hampered by the violent forces they unleash, which can destroy people and instruments. A few aeroplanes, however, have managed to ride a hurricane, their crews surviving to describe the inside view. The heaviest rain, they say, looks as if it is falling in spiral bands, and super-cooled water vapour turns instantly to ice on the aeroplane's wings. Within the eye of the hurricane, the air is pure and clear, the surrounding banks of cloud rising like the tiered galleries of a grand opera house to a dome of clear blue sky far above.

TAMING A HURRICANE

Stormfury was the name American scientists gave to a programme that, they believed, would snuff out hurricanes before they could wreak serious damage. The plan was to 'seed' the clouds of a hurricane with crystals. Moisture would gather around the crystals to produce droplets big enough to fall as rain. Inducing the hurricane cloud to release itself as rain would enlarge the hurricane's eye and so reduce the intensity of the circulation.

It did not work. On a trial run with an Atlantic hurricane, the winds did drop, but picked up again almost immediately afterwards. The hurricane then changed its track and went on to hit Savannah, Georgia. US Navy weather experts were never sure whether the change of course was due to this interference, and it led them to end the programme in 1971. Plans for more experiments in the Pacific were also abandoned when Japan and China objected.

Ever Watchful *A US Navy photographic plane keeps watch on Hurricane Gracie (1959); the hurricane's eye is above the plane's tail.*

Aircraft called hurricane hunters, which go around the edges of storms, are equipped with special radar systems (capable of detecting the speed of a single raindrop), with which they track the position and direction of storms. Satellite instruments, meanwhile, capture aspects of a hurricane that would be impossible from any Earth-based source, including all-in-one views of the mature system. Sensors register temperature changes in the ocean surface, and produce coloured satellite photographs that reveal the trail of cold water following in the wake of a hurricane, where heat has been drawn up to fuel its progress. Most important of all, the satellites make constant observation possible and keep track of every seedling, or small cyclonic depression, that might develop into a full-blown hurricane.

In 1982, one satellite brought to the attention of meteorologists a typhoon that might otherwise have gone unnoticed. Early detection of its sudden change of course towards Hawaii meant that advance warning could be given. When the typhoon hit Hawaii 24 hours later, vulnerable areas had been evacuated and, although damage amounted to around $200 million, only one person died.

Official US hurricane centres – in Florida (covering the Gulf of Mexico and the Caribbean), Puerto Rico (Atlantic), San Francisco (eastern Pacific) and Honolulu (central Pacific) – keep target areas informed and issue two levels of bulletin. Hurricane Watch alerts stretches of coast if there is a 50 per cent chance of a hurricane within 36 hours; and Hurricane Warning when a hurricane is about to strike.

Viewed From Afar *Doppler radar is used to determine wind speeds in a hurricane.*

Observation Flight *Modified and strengthened aircraft are used to study conditions within thunderstorms.*

Aftermath *Powerful winds, generated in 1992 by Hurricane Andrew, sheared off the side of a building in Miami (right).*

TORNADO TERROR

A spiral of fast wind, a tornado can shatter anything in its path, as it cuts across the open landscape. Although only a few hundred feet in diameter, and lasting no more than an hour or two, its thin trail of destruction can run for miles.

In late April 1991, a cold mass of air was moving south-east from Canada across the Midwestern United States. It reached the Rocky Mountains and veered south. Simultaneously, warm moist air was moving up from the Gulf of Mexico. The two air masses met in Kansas on April 26.

The warm air lay beneath the cold dry air, separated by a layer of stable air. The warm air then burst through the layers above. This uprush resulted in a sudden drop in pressure, causing the moist air to condense into droplets and form thick storm clouds. Lightning flickered through the darkness as the turbulent air swirled and eddied.

Suddenly, just before 5 pm, near Argonia, Kansas, a tube like an elephant's trunk descended from the base of the cloud towards the ground. It vanished after a few minutes. Then, a few miles farther east, a fully grown tornado appeared and drilled into the town of Clearwater. It screamed into the houses on the edge of the town, lifting roofs and picking up cars as if they were toys, swallowing and pulverising everything in its path. Wooden houses were torn from their foundations; burst water mains fountained into the air, and power lines and poles were torn apart.

The 'twister' – the local term for a tornado – swung east to weave among the fighter planes, residences and school of a nearby airbase. It increased in intensity over the nextdoor town of Andover, sometimes splitting into three vortices and screaming like ascending jet planes. Emergency services had been alerted, but there was no time to organise an evacuation. When the funnel, swollen with the debris of its destructive course, hit the southern part of Andover at 5.40 pm, 84 houses and many shops and offices simply vanished.

A policeman drove to alert the families living in the Golden Spur Mobile Home Park, but they had already seen the still-growing tornado. Over 200 managed to reach the safety of the park's custom-designed tornado shelter – their flimsy homes were tossed and smashed to splinters and sucked away. Several hundred homes in Golden Spur were destroyed, and 13 of their owners lay dead.

TUBE OF DESTRUCTION

Tornadoes are born from a titanic struggle between unstable cool, dry air lying on top of warm humid air, with a thin lid of stable air between. The warm

VICIOUS TWISTER *Tornadoes can form a winding tunnel of dark dust over the American Midwest (left), or a waterspout over seas and lakes (right).*

air can sometimes burst up through the lid, sweeping up powerful winds into a vicious thunderstorm called a supercell. The supercell starts rotating – anticlockwise in the Northern Hemisphere and clockwise in the Southern. Known as the Coriolis effect, this is a result of the Earth's rotation on the movement of the atmosphere. As the supercell rotates, it pulls up cold air spilling down from the edge of the storm and warm air from the ground. Where the two meet at the centre of the supercell's rotation, they entwine and spin upwards into a funnel, spinning faster as they rise. The air is usually moist and the drop in pressure at the centre of the storm condenses it into droplets giving the tube an opaque, white appearance. This turns a dirtier colour as the tornado vacuums up the debris in its path.

Tornadoes in the United States are most likely to occur from March to August when the differences between cold and hot masses of air are greatest, and early in the afternoon when ground temperatures are at their peak to fuel further convection currents.

Tornado Alley is the name popularly given to a zone of land stretching from the Gulf of Mexico, north through the states of Texas, Arkansas and Oklahoma,

WARNING SIGNS *Thick clouds with bulbous underbellies (left), building up along the line of a front (far left), may indicate tornado conditions.*

ANATOMY OF A TORNADO

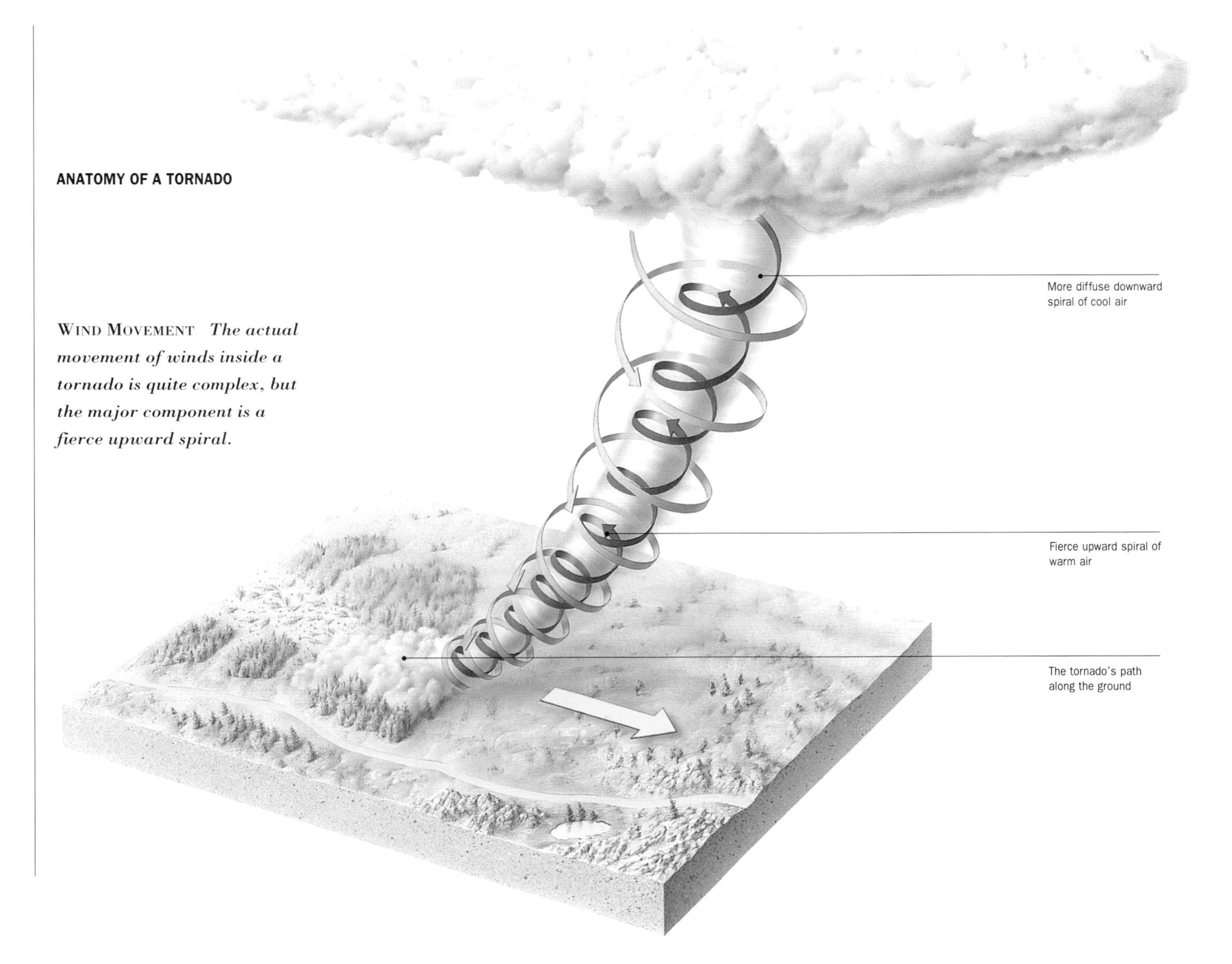

WIND MOVEMENT *The actual movement of winds inside a tornado is quite complex, but the major component is a fierce upward spiral.*

WEST AFRICAN VARIATION

In West Africa, sudden, short squalls from severe thunderstorms are described as 'tornadoes'. Characterised by torrential rain, these may have fronts up to 200 miles (320 km) long. Usually, they happen between the wet and dry seasons, and are caused by the meeting of warm, wet monsoon air from the south-west with the dry north-eastern *harmattan* wind from the Sahara Desert.

TORNADO ALLEY *Dusty tornadoes, such as this one in Nebraska, are common throughout the region that has come to be known as Tornado Alley.*

Kansas, Nebraska and the Dakotas, as far as Canada. Here, where cold, dry air masses from Canada confront warm, wet air from the Gulf of Mexico, around 700 tornadoes occur each year.

ALLEY OF DEATH

Sometimes a whole string of them forms, one immediately after the other. Around noon on April 3, 1974, a tornado appeared on the horizon of Illinois state.

Over the next 21 hours, 148 ripped through 13 states from Alabama in the deep South to Ontario, Canada, all of them in or bordering Tornado Alley. The death toll was 315, with more than 6000 injured and 28 000 families left homeless. The tornado epidemic became known as the '1974 Super-Outbreak'.

Some people have actually survived the experience of a tornado passing over them. They describe an earsplitting, hissing scream, similar to the noise when a jet plane or express train passes close by. There are sometimes flickers and flashes of lightning, and a strong smell of ozone that is reminiscent of rotting seaweed.

Scientists have still not determined exactly why tornadoes happen and how they work, partly because every one is different, and partly because instruments laid out in the path of a tornado rarely survive. However, calculations estimate that the spiralling winds inside the tornado can reach speeds of 174-249 mph (280-400 km/h) and sometimes as much as 311 mph (500 km/h). Tornadoes can come and go within minutes or even seconds, and none is known to have lasted for longer than 3 1/2 hours. Most tornado paths are 150 ft (46 m) wide, but very destructive ones may be over a mile (1.6 km) wide. The trail of damage that is left in their wake is, on average, 148 ft (45 m) wide. The speed at which they move across the ground can be up to 75 mph (120 km/h), but the distance covered may only be a mile or two (less than 3 km) before they disappear. Only a very few – about 0.5 per cent – travel more than 100 miles (160 km).

THE TRI-STATE TORNADO

The cone-shaped 'Tri-State Tornado' of March 1925 was one of the most destructive recorded in North America. At about 1 pm, a huge thundercloud built up over Ellington in Missouri. Fifteen minutes later, a broad vortex appeared out of it and bored into the ground, flattening the town of nearby Annapolis.

The tornado then turned and headed north-eastwards at an unusually fast 60 mph (96 km/h), blasting through the town of Gorham in neighbouring Illinois. Its base

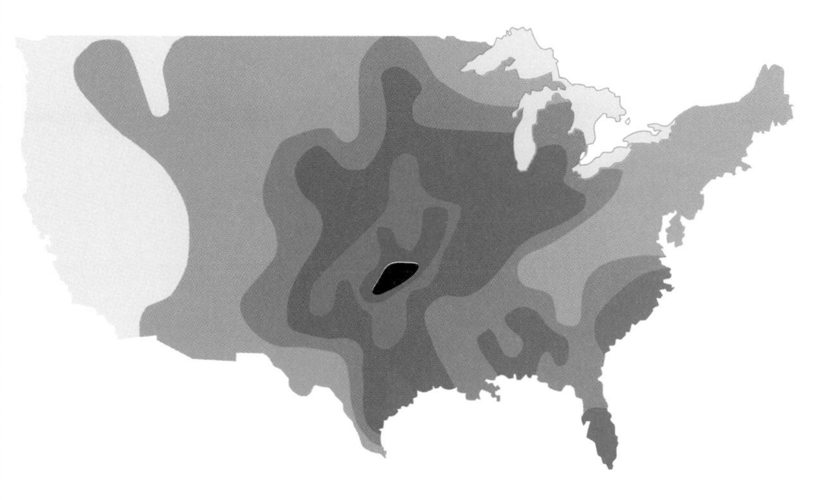

YEARLY OCCURRENCE OF TORNADOES IN THE UNITED STATES

DANGER ZONE *Tornadoes are common along a line that stretches up through the United States.*

APPROACHING TERROR *The dark funnel of a tornado against the lowering sky (above) may be difficult to see at first, but the noise is unmistakable.*

broadened from 1/4 mile (400 m) – already unusually large for a tornado – to about a mile (1.6 km) as it crossed to Indiana, picking up people like pieces of rag and putting them down 1 1/2 miles (2.4 km) away. It destroyed the town of Griffin and part of nearby Princeton before fading away. In three and a half hours, the Tri-State Tornado had travelled 219 miles (350 km), leaving 695 people dead and more than 2000 injured. The same storm triggered smaller

DOUBLE TROUBLE *On rare occasions, particularly fierce conditions produce a tornado with twin funnels (right), as in Indiana in 1965.*

See a Bug...
Call
Arab
TERMITE & PEST CONTROL

A SENSE OF PROPORTION

The following scale of tornado power, based on maximum wind velocity, has been devised in Japan by Dr T. Fujita.

Level	metres/second	km/h	mph
0	18	65	40
1	33	120	74
2	50	180	112
3	70	252	156
4	93	335	208
5	117	421	262

tornadoes in the neighbouring states of Kentucky and Tennessee, claiming a further 100 lives.

TRACKING A TORNADO

Tornadoes are notoriously difficult to predict. Given the right conditions, the updraughts of warm air that create the thunderclouds are given a rapid spin and turn into a fully fledged tornado. But if all of the world's tornadoes start in thunderclouds, it is not the case that all thunderclouds produce tornadoes. Even in Tornado Alley, only 1 per cent of all thunderclouds give birth to them.

The danger signs are when the base of a thundercloud becomes hummocky from the churning conditions within. Dark patches of cloud hang from the base of the stormcloud, looking like clusters of cows' udders – hence the official term is *mammata* (festooned) cloud.

Since 1977, Doppler radar has greatly improved tornado forecasting – from 69 per cent to 94 per cent accuracy – by detecting the telltale circular movements of supercell thunderstorms. The radar can now predict tornadoes about 25 minutes before they emerge from the base of the clouds – sufficient time to issue emergency warnings – and has helped to reduce the number of casualties from tornadoes in the USA.

Another project, based at the National Severe Storms Laboratory in Oklahoma University, is making more accurate studies by dropping instruments in and around tornadoes. An armada of trucks have weather stations mounted on their roofs and collect soundings from high-altitude balloons. Others carry armoured packages of instruments which are strong enough to withstand the tornado. These are placed on the ground to measure temperature and pressure as the tornado passes. This project, called VORTEX (Verification of the Origin of Rotation in Tornadoes Experiment), has been in operation in Tornado Alley since 1994.

On a less high-tech level, there are people living in the states that straddle Tornado Alley who describe themselves as 'tornado

DANGER SIGN *Hummocky mammatus clouds warn of an imminent tornado. Each lump is a cell of circulating air that may give rise to a whirlwind.*

COUNTING THE COST *Although the destructive track of a tornado is quite narrow (top), the damage is intense (right).*

TORNADO RUINS *Pressure differences cause houses to burst outwards as a tornado passes.*

chasers'. Following the paths of tornadoes by car, filming their progress and plotting their routes, these amateur meteorologists use CB radio to give advance warning to towns that seem to be most immediately vulnerable.

WHEN THE POWER IS RELEASED

Another ominous sign can be picked up in satellite photographs – when a swelling appears above a thundercloud's characteristic anvil shape, where it has frozen and spread out in gauze-like cirrus clouds. Areas with intense convection currents can also create broadly circular clouds which form a long, curved vertical wall ending in a sinister hook. Such conditions can cause a tornado.

Bark flies from the trunks of trees, people are stripped of their clothing, sheep are shorn of their wool, and chickens are plucked of their feathers – the sudden and enormously concentrated force of tornadoes has been responsible on different occasions for all these effects.

The difference in pressure between a tornado's core and the surrounding atmosphere can be as much as 100 millibars – a near-vertical pressure gradient over a matter of a few yards. But most of the damage is caused by the brute force of the wind, rather than by the sudden drop in atmospheric pressure in the tornado. Home-owners were once advised to open all doors and windows to reduce the pressure inside a building. Now residents are urged to keep everything shut and to shelter in a central room, where they will be best protected by internal walls. The conflicting messages reflect how little experts still understand the phenomenon.

The main destructive element of the tornado is the violent shearing force of the funnel's whirling winds, slicing buildings open like a saw, yet sometimes leaving the contents completely untouched. But the debris caught up in the funnel can be flung at fantastic speeds – straw can be fired like bullets into tree trunks and wooden planks into iron girders. Heavy items, such as pieces of roof, can be hurled tens of miles and lighter debris like paper can be scattered up to 200 miles (320 km) away.

A gust of wind suddenly blasts around

continued on page 108

FREAK SHOWERS

Throughout history there have been reports of showers of fishes, frogs, rats, lizards and other unlikely items. Some can be put down to superstition, but many may be the result of a tornado's suction and dumping powers. In 1978 a flock of geese in Norfolk, England, was picked up by a tornado and later dropped out of the clouds over a 28 mile (45 km) area.

FUNNEL OF FURY *A tornado (overleaf) passes over Texas on May 13, 1987.*

TOP TORNADOES

Year	Location	Comments	Deaths
1903	*Georgia*	*Mill collapsed*	*98*
1925	*Missouri, Illinois, Indiana*	*Tri-State Tornado*	*695*
1953	*Massachusetts*	*Worst in New England*	*94*
1965	*Mid-West*	*37 tornadoes in one day*	*271*
1970	*Oklahoma*	*Warning and evacuation worked*	*0*
1974	*Alabama to Ontario*	*The Super-Outbreak – 148 in 21 hours*	*315*
1991	*Kansas and Oklahoma*	*Mobile home park destroyed*	*29*

the corner of a building, swirls in an eddy and shoots skywards, funnelled round the buildings and propelled by convection currents from hot streets. It picks up cigarette butts, candy wrappers and flurries of dust, spinning and tossing them upwards.

Poor Relations

Dust devils or whirlwinds are usually smaller spirals of spinning air than their bigger brothers, tornadoes. But unlike tornadoes, they are created simply by hot ground heating the air above into a rotating funnel.

A tornado over water produces another variation – a waterspout which descends from its parent thundercloud in a spinning cone. The water becomes severely agitated before the point of the cone reaches its surface, so that the descending point is ringed by a cloud of spray. The waterspout itself is white, as water droplets are whirled upwards. Accompanied by its surface corona of spray, the spout moves over the water like a column that becomes looped or slanted if the top moves at a different speed from the base. Waterspouts last between ten minutes and half an hour, and are usually less energetic than their land-bound counterparts, although some are capable of picking up and wrecking boats and jetties.

A tornado that formed at Norfolk in Virginia in 1935 started to demolish the town, crossed a creek sucking up water until the bottom of the creek was exposed, and then gouged a channel in the mud. As a waterspout, it lifted small boats on to the shore, ripped off part of a heavy pier and then went back on to the land again where it destroyed several buildings.

True waterspouts can also cause considerable damage. The White Star liner *Pittsburgh* was hit by a huge waterspout in the mid-Atlantic in 1923. So many tons of water were dumped on her so suddenly that her bridge was wrecked.

Sailing through a waterspout can be a terrifying experience, as John Caldwell described in his book *Desperate Voyage*. He was sailing his boat *Pagan* in the mid-Pacific when he saw a waterspout and deliberately headed for it: '*Pagan* was swallowed by a cold wet fog and whirring wind. The decks tilted. A volley of spray swept across the decks. The rigging howled. Suddenly it was dark as night. My hair whipped my eyes, I breathed wet air, and the cold wind wet me through. *Pagan*'s gunwhales were under and she pitched into the choppy seaway.

'I had sailed into a high dark column from 75 to 100 ft [22.5-30 m] wide, inside of which was a damp circular wind of 30 knots, if it was that strong.

'As suddenly as I had entered the waterspout I rode out into bright free air. The high dark wall of singing wind ran away.'

Devils and Spouts *Dust devils (below) and waterspouts (bottom) are differently coloured because of the different substances sucked up in them.*

Tornado at Sea *A waterspout appears as a waving column of water spray off the Spanish coast in 1965 (above and right).*

FLOOD ALERT

Heavy storms may sometimes cause huge volumes of water to depart from a river's natural course and spread over previously dry land. Unstoppable and all-pervasive, the resulting floods can be devastating to human life.

The city of Florence in northern Italy sits on the banks of the Arno 55 miles (88 km) from its mouth, just where the river channels out of the Apennine mountains and onto the coastal plain. From Roman times, its position has made it a centre of commerce, and by the Renaissance it had become the cultural centre of Europe. However, its site has also made it very vulnerable to flooding, as noted by Leonardo da Vinci. Between 1500 and 1510, Leonardo sketched a plan for engineering works that would lessen the threat. These involved diverting the Arno by means of a canal and tunnel, and establishing a vast basin to collect flood water. His advice was ignored and the scheme forgotten, as the situation continued to worsen. Over the centuries, the trees on the forested slopes of the Apennines, upstream from Florence, were cut down and never replaced. The soil lost its ability to retain water, and every rainstorm brought water that ran straight off into the Arno instead of lingering in the soil and vegetation.

On November 3, 1966, the area suffered one of its worst ever storms. About one-third of the annual expected rainfall was delivered in 40 hours of hard and steady rain. Run-off from the bare hills filled the mountain streams and these swelled the River Arno itself, as it roared down towards the city, which was settling down for the night. Throughout the evening, the river's channel through the town became deeper and deeper until, by midnight, it had risen 17 ft (5 m) and was racing onwards at a speed of 80 mph (130 km/h). The Arno's defences were ruptured, and the dark mountain waters gushed into the lower-lying areas of the city in a chaos of spiralling whirlpools, hurling debris about like battering rams and sucking helpless people down in backwashes and countercurrents.

The stricken city was plunged into black night when underground transformers short-circuited. In the darkness, cars were picked up and swirled about like old shoes. Oil tanks were ripped from their mountings and ruptured, spreading oil on the flood waters and turning them into a muddy, corrosive slime. The city's sewers were put out of action, the sudden influx of flood water building up pressure and sending human excrement fountaining into the air from a hundred manholes.

The Bargello and Santa Croce districts to the north of the river, a densely populated area of narrow alleys and little workshops, were particularly badly hit. Here, the populace was caught unawares and found it difficult to flee as the water thundered through the winding streets. Thirty-two people perished in Florence, and a further 112 along the Arno river

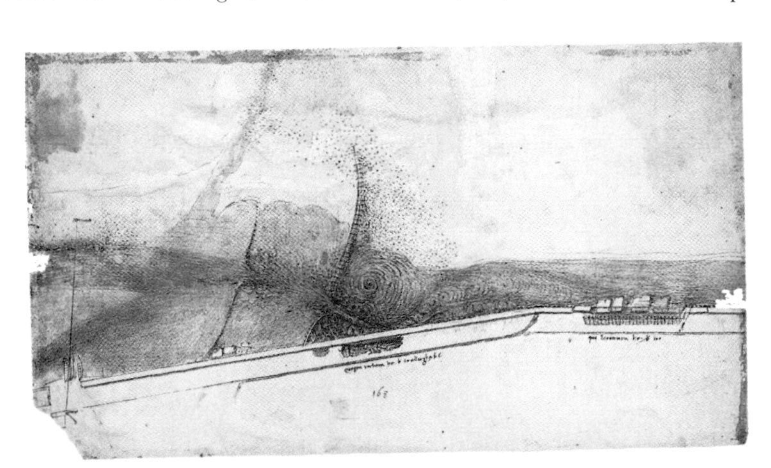

ANTICIPATION *The danger posed by the River Arno was appreciated in the 16th century, when Leonardo da Vinci drew up a flood plan.*

TWO EXTREMES *Even arid landscapes, like those in western New South Wales, Australia, can be subjected to disastrous floods (right).*

CHAOS *Debris and mud caused irreparable damage in the Florence flood of 1966.*

FLOOD DAMAGE *Manuscripts from Florence's National Library are laid out to dry. Part of Paolo Uccello's* Scenes from the Life of Noah *was among the damaged paintings.*

basin, 800 communities and some 10 000 houses were affected, 16,000 agricultural machines were destroyed and 50,000 farm animals were either killed or had to be slaughtered.

Another tragedy was the loss of many of the city's cultural treasures. The bronze panels sculpted by Ghiberti for the doors of the cathedral Baptistery in the Piazza del Duomo were wrenched off and swept away. The basements of the museums were filled with filthy water. More than 1.5 million books stored in the National Library were saturated with the slime, two-thirds of them beyond repair. In the Uffizi Gallery 130 000 photographic negatives of the world's greatest paintings were lost. What was damaged took years to restore.

Hundreds of art students from all over the world were studying in Florence at the time, and they played a decisive role in the rescue. These young people plunged into the stinking pools to rescue all that could be carried, forming a human chain that reached down into the flooded basement of the National Library to bring out the manuscripts and other treasures before they were lost forever. The Archaeological Museum was choked with mud, and it was remarked – ironically – that the rescuers had to resort to a whole new archaeological excavation in order to extract the artefacts that were kept there. Paintings, some of them 20 ft (6 m) tall and 10 ft (3 m) wide were stored to await restoration, each with a plastic bag attached to contain the paint flakes that had peeled off the buckling, sodden canvas.

Art restorers from all over the world converged on the sad city, and every means available was used to restrict the damage. The leaf-drying sheds of local tobacco firms were commandeered for drying priceless

AFTER SEVEN CENTURIES *Water damage is obvious on Giovanni Cimabue's* Crucifixion *seen here before (right) and after (left) the Florence flood.*

BENIGN FLOODS *The fertile strips along the banks of the Nile are caused by annual flooding.*

historical documents. Thirty years later, two-thirds of the 3000 paintings damaged are restored and back on display, and the rescued artefacts are now stored well above the likely water level of any future flood. The restoration effort that followed the flood actually boosted the science of art restoration, and techniques were developed that are used throughout the world today; the Italian Ministry of Culture now runs a laboratory in Florence that is the world centre for restoration techniques. There are now more art treasures on display in Florence than before the flood, and these are in a better state of repair than ever before.

WATER EVERYWHERE

The flood that devastated Florence was the result of excessive rain, one of the major causes of flooding – but not the only one. Others include an abnormal rise in sea level brought about by a combination of tides and wind. Low-lying countries are the worst affected by this phenomenon. Bangladesh, for instance, sited on the deltas of the Ganges and the Brahmaputra rivers, and nowhere more than a few yards above average sea level, is frequently afflicted by typhoon-driven floods. Dam bursts are another cause of devastating floods, but these tend to be man-made rather than natural disasters.

Flooding by rivers swollen by heavy rainfall is not always a disaster, however, for when it is controlled it can form the basis of the agricultural economy of an area. In Ancient Egypt, the annual flooding of the Nile was welcomed, with the excess water being diverted into irrigation systems essential to the agriculture of such an arid area, and bringing much-needed fertile silt to add to the local soil. On a smaller scale, the water meadows of Britain, which date from the early days of agriculture, were allowed to flood during the winter, replenishing the soil with material brought down from the surrounding uplands.

Perhaps the region most prone to flooding from rivers swollen by precipitation is the flood plain crossed by China's great rivers. In 1887 the Huang He – sometimes called the River of Sorrow – burst its banks killing more than a million people. Chinese farmers, using nothing but logs, stones and soil, had to repair a hole in the bank several hundred yards wide. They succeeded but it took them over a year. Then, in 1931, heavy rains triggered six devastating flood waves on the Yangtze River, destroying 23 dams and dykes. Forty million people were left homeless and 3-4 million drowned as a result of cyclone flooding and the famine which followed.

During great disasters like this, the rivers usually change their courses as well, bringing profound changes to local society. The earliest recorded flood disaster on the Huang He was in 2297 BC. Shortly afterwards, the emperor Yu the Great established a series of waterworks on China's flood plains to try to keep such events in check, but the system deteriorated during times of war and of corrupt rule.

Since records were kept, the Huang He has changed its course several times and over vast areas. It used to enter the Po Sea to the north of the Shantung Peninsula. Then, in AD 1289, during one of its great floods, it flowed southwards and became a tributary of the Yangtze, reaching the sea south of the Shantung peninsula, some 600 miles (965 km) south of its original mouth. Then in 1855, during another flood, it resumed its original course and entered the Po Sea once more, as it does today. At times of severe flooding it actually reached the

OFT-REPEATED CHORE *As the Yangtze River rises (top), villagers on the lower reaches strengthen the flood barriers.*

sea both north and south of Shantung, turning the peninsula into an island. Human influence has not always been for the best. In 1938 the Japanese army invaded China. The Chinese, in an attempt to stop them, breached the levees of the Huang He, hoping that the resulting flood would halt the advance. It did not – but the floods killed an estimated 1 million Chinese.

Old Man River

In the Western world it is the Mississippi River, which drains about a third of the land area of the United States, taking water from all the great mountain ranges of the continent and delivering it into the Gulf of Mexico at a rate of 176 million million cu ft (5 million million m^3) per year, that commands the attention of flood engineers. Much of its lower reach consists of flood plains, and the city of New Orleans, founded in 1727, has been flooded regularly. It was only in the mid 19th century that flood protection began to be taken seriously by the federal government, and only in 1928 that the problem was addressed scientifically. At Vicksburg there is a column, built into a flood defence wall, that shows the height above normal of the Mississippi during times of severe flooding. The topmost mark bears the ominous inscription: '62.3 feet [18.9 m], if levees had held, 1927'.

In fact, the flood of that year threatened New Orleans itself, and levees had to be breached with explosives to draw the water away from the dangerous areas. This followed an unprecedented period of rainfall that had started the previous August. On April 19, the levees at Greenville, Tennessee, began to break. Farmers and townspeople rushed to help, shoring up the banks and strengthening them with sandbags, but the levee collapsed while they were working it. The tributaries of the Mississippi reversed their flow, backing up with the height of the main river. At Pine Bluff on the Arkansas River, about 60 miles (95 km) from where it joins the Mississippi, 500 people crowded onto the bridge to watch the flood waters. The waters rose farther than they expected, and all 500 were stranded for three days (during which time two babies were born on the bridge). When New Orleans itself was threatened, the great Caernarvon levee to the south of the river was breached with dynamite to

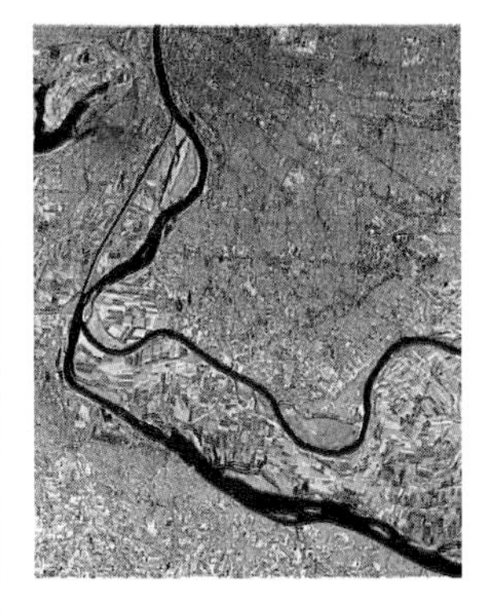

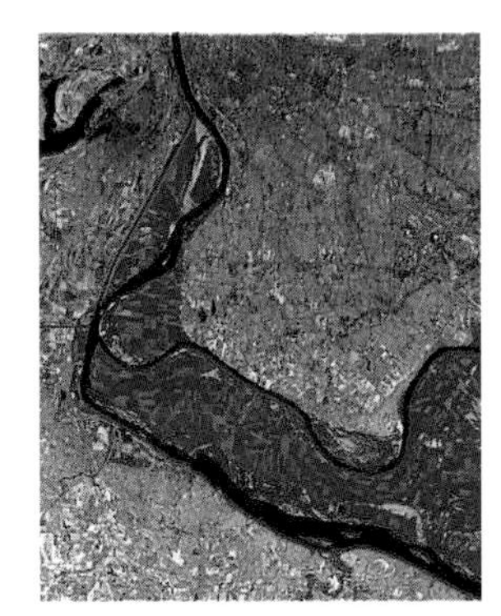

Red Alert *Satellite photographs show St Louis before (left) and during (right) the Mississippi floods of 1993. Floodwaters show up red.*

Inundation *The rooftops of an isolated farm, a victim of the Mississippi floods in 1993, are all that remain visible.*

BROKEN DEFENCES *Thick brown flood waters stretched as far as the eye could see when the Mississippi breached its levees (far right) in 1993.*

lower the level of the water. The levee was so well built, however, that it took several attempts by engineers to break it.

A national rescue appeal went out to help people across the affected area in the states of Mississippi, Arkansas and Louisiana, and 6000 river boats were volunteered; local sawmills turned out 1000 new boats in ten days; and a further unexpected source of help came from local bootleggers around Greenville. Equipped with fast, manoeuvrable boats and formidable knowledge of the local area, they played a big part in the rescue operation. That year 246 people died, 650 000 were made homeless and much of the area was submerged in a vast lake 80 miles (130 km) across.

Possibly the most spectacular flood caused by heavy but brief periods of rainfall is the 'flash flood'. Water falling on the catchment area of a stream is channelled, via surface run-off and tributary streams, into the main stream itself, producing a sudden filling of the stream, followed by just as sudden a return to its usual level. Flash floods occur most dramatically in desert areas. Infrequent but heavy rainstorms

A FLOODED WORLD

Warming of the planet has the potential to cause extensive flooding across the world but predicting how much, when and where is difficult. A temperature rise of 1°C (1.8°F) would theoretically raise the sea level by 2 ft (61 cm), as a result of the sea water expanding as it heated up, regardless of the effect of any melting ice.

However, if global warming continues, more water will evaporate from the surface of the oceans, resulting in an increase in precipitation in high latitudes, more snow and ice, and therefore more water locked up on land. Between the late 1970s and the early 1980s, satellite measurements showed that the Greenland ice cap was becoming thicker at a rate of some 9 in (23 cm) per year as a result of evaporation from the oceans due to higher temperatures. This corresponds to a lowering of the worldwide sea level by about 1/10 in (2 mm) each year. A temperature rise of 3.5°C (6.3°F) would probably increase the volume of water locked up as ice around the world by 40 cu miles (100 km^3). However, if the temperatures rose by 10°C (18°F) and remained there, the ice caps would melt more quickly than they would accumulate, and sea levels would rise dramatically.

ICE CAP *The ice in Antartica may not be permanent.*

Ice that is floating on the sea, in the form of icebergs, the ice shelves of Antarctica, and the North Polar ice cap, displaces roughly its own volume of water: as it melts, it will produce little net difference in sea level. The greatest difference will be caused by melting land-based continental ice sheets. The Greenland ice sheet would contribute 26 ft (8 m) to the rise in sea level; the West Antarctic ice sheet would add 20 ft (6 m); the melting of the East Antarctic ice sheet, the smaller ice sheets such as those in Iceland, and all the valley glaciers of the world, would bring the sea level to about 230 ft (70 m) above present levels. A rise of a mere 17-35 ft (5-10.5 m) would submerge a land area as great as China. The speed of such an inundation is impossible to estimate. Research has shown that during the Ice Age there was a period, about 95 000 years ago, when the sea level rose by 40-50 ft (12-15 m) within a century.

Add to this the fact that many of the landmasses of the Northern Hemisphere are tilting, and the calculations become really difficult.

Once the glaciers had retreated after the last ice age, land that had previously lain beneath them was freed from their weight and sprang back into position. This rebound is still taking place, with the result that the north-west corner of North America is rising by a rate of about 3/4 in (20 mm) per year, and the north-west coast of the British Isles is rising by about 1/2 in (12 mm) per year. However, the opposite corners are sinking: Louisiana and Texas at a rate of 1/3-1/2 in (9-12 mm) per year, and south-east England at a rate of about 1/10 in (2 mm) per year. These are the areas that are at the greatest risk of inundation by the sea.

The Unexpected *A flash flood hurtles through a normally dry canyon in Arches National Park, Utah.*

in such areas, where the ground has been compacted by previous bursts of heavy rainfall and is therefore impermeable as concrete, producing high run-off, can cause a canyon or an arroyo that is dry for the rest of the year to fill suddenly.

One of the most destructive flash floods on record took place in 1976 on the Big Thompson River in Colorado. Normally this river is only 18 in (46 cm) deep, and just a few feet wide where it joins the South Platte River. The Big Thompson Canyon, through which the river flows for 25 miles (40 km) before crossing the plain, is a noted beauty spot. On July 31, 1976, the area's population of 600 had swollen to 3500 as people enjoyed a holiday to celebrate Colorado's centennial. For a day or two before, thunderstorms had been building up over Kansas and Colorado, but the rains started in earnest at about 6.30 pm.

Over the next four-and-a-half hours, 12 in (30 cm) of rain fell, exceeding the average annual rainfall of the region. The weather services had known that this was coming, but the rescue authorities were paying all their attention to the Mississippi, for fear of floods there. The local meteorological station in Denver was closed down for repairs. And, as a result, the holidaymakers in Big Thompson Canyon had no warning of what was about to happen.

Witnesses state that the approach of the flood down the canyon sounded like oncoming freight trains. A wall of water thundered down the gorge, thrusting a cloud of dust before it, so that people had no clear idea of what was coming. It burst into the landscape at a speed of 20 ft (6 m) per second, emptying 233 000 gallons (1 million litres) onto the plain every second. Police cars racing to help were engulfed as US Highway 34 was ripped up, 10-12 ft (3-3.7 m) chunks of asphalt surface hurtling into the air, as the road washed away. That night, the 18 in (46 cm) deep river reached a depth of 32 ft (10 m) resulting in 139 deaths by drowning.

MAJOR RIVER FLOODS

Date	Country	River	Casualties
1824	*Russia*	*Neva*	*10 000*
1887	*China*	*Huang He*	*1 million*
1913	*USA*	*Ohio*	*500*
1931	*China*	*Yangtze*	*1 million*
1933	*China*	*Huang He*	*3 million*
1938	*China*	*Huang He*	*3-4 million*
1948	*China*	*Min*	*3500*
1954	*China*	*Yangtze*	*30 000*
1988	*Bangladesh*	*Ganges and Brahmaputra*	*1000*
1993	*USA*	*Mississippi*	*48*
1997	*USA*	*Red*	*Unknown*

The Unpredictable Sea

Rainfall and rivers are one source of flooding; wind and sea provide another. On January 29, 1953, a depression formed off south-west Iceland and began to move in a

HOW TO PREVENT COASTAL FLOODING

The most effective barrier against the erosion of the world's coastline is a natural beach or salt marsh. But although most of the beaches of the world were established with the retreat of the glaciers at the end of the Ice Age, they are quite transient phenomena. With the growth of leisure time in manufacturing nations during the 19th century, beaches became recreational sites. The concept of a 'holiday resort' was established, and many of these were centred on areas with extensive beaches, usually within reach of large centres of population. In the short timescale of human economics, a beach seemed to be a permanent resource.

However, over the decades, the natural forces of erosion proceeded to change the nature of the world's beaches, and this was seen as an economic disaster. All kinds of measures were introduced to prevent beaches from being washed away, including the emplacement of groynes – heavy fences built out across the beach at right angles to the shoreline to prevent sand from drifting along the shore. These interrupted the natural evolution of those parts of the coastline, however, depriving any beach farther along the coast of its sand, with the result that the land behind began to erode.

Most beach sand is derived from the erosion of nearby cliffs, but if there are buildings on top of these cliffs, people tend to take preventive action by facing the foot of the cliff with concrete to reinforce it, or by building up breakwaters so that the erosive power of the waves is checked before it reaches the cliff. The result may halt the erosion of the cliff, but it will deprive the local beaches of their new sand, contributing to a problem that has become international. It will be difficult to persuade British landowners, for example, to allow their cliffs to erode in order that the coastline of the Netherlands will be well supplied with sand for their coastal protection.

Modern thinking, therefore, tends to promote the development of natural beaches or salt marshes, with any sea walls to the landward side of these natural barriers. This involves establishing a line of defence well back from the sea. The major problem will be to persuade people that what looks like a retreat is, in fact, the correct and natural approach.

GROYNES *Wooden fences built into the sea prevent shingle beaches being washed away.*

STEPPED SEA WALLS *The wall's profile breaks the power of the waves and guards against coastal erosion.*

south-easterly direction. Two days later, the pressure had fallen to 968 millibars – very low for the area. The first tragedy of the freak weather – caused by the combination of a deep depression, gale-force winds and the arrival of the spring tide – was the sinking in the Irish Sea of the car ferry *Princess Victoria* with the loss of 132 lives. Severe winds spiralling into this depression drew water from the southern North Sea as the area of low pressure crossed the North Sea towards Denmark. This depression was immediately followed by a ridge of high pressure that swept the excess water before it. Sea levels everywhere along the east coast of England south of the Humber estuary were raised by over 6 ft (2 m), rising to over 10 ft (3 m) along the Dutch coast. By the early morning of February 1, the surge of water had flowed around the basin of the North Sea and was sweeping out into the open ocean again, past Denmark and Norway.

In the wake of this surge came chaos and devastation. The sea had burst through the sand dunes that had been protecting the land for 300 years and had flooded the low East Anglian landscape. In areas where the sea walls had kept back flooding since historical times, the abnormal height of the water generated by the storm combined with the high spring tide and the 119 mph (190 km/h) winds to spill over the defences. Water poured down the rear of the walls, gouging out the foundations on the unprotected side and causing whole lengths to collapse. It gushed up the canals of the fenlands, built originally so that the water would drain away from them. A wall of water 8 ft (2.4 m) high swept inland engulfing villages 5 miles (8 km) from the sea. The insubstantial wooden buildings of a bungalow park near Hunstanton collapsed in the onrush, killing 65 of the inhabitants. The 7.27 train from Hunstanton to King's Lynn, battling through the rapidly rising

(continued on page 120)

RESCUE *Troops and police evacuate families trapped in their homes by rising sea water in King's Lynn, Norfolk, 1953.*

The Flood-plain River

Rivers are traditionally regarded as having three stages. In the first, the water comes tumbling off the mountains, picking up debris and eroding its bed and banks as it goes. In the second stage, it travels through broad valleys and slows down. The slower a current flows, the less material it can transport, with the result that, during this second stage, it deposits debris as well as gathering it up through erosion. In the third and final stage – the flood-plain stage – very little erosion takes place and much of the material carried in suspension by the current is deposited to form the flood plain.

During a flood, a river overtops its banks. The resulting overflow gradually slows in momentum as it spreads away from the river, and as it slows, it deposits large quantities of the material it is carrying. This creates a great build-up of sediment along the banks of the river, which become higher and higher with every flood and form natural features known as levees. All the while, sediments are also accumulating on the floor of the river itself – to the point where the surface of the river, and even its bed, are higher than the surrounding plain, and are held back only by the levees. It is a very precarious balancing act. The surface of the Huang He is as much as 33 ft (10 m) above the level of the flood plain. If the levee is breached for any reason, the surrounding land floods very quickly.

A river winds across its flood plain in vast loops, called meanders. On the outside of the loop the current is fastest, and the outside bank is continually being eroded away. On the inside of the loop the current is slower, and so suspended material tends to be deposited here. As a result, the loop tends to become bigger as time goes on until, eventually, the neck of the loop becomes so narrow that, during a flood, it can be cut off as the river follows a new, shorter course, abandoning the loop as an oxbow lake. The oxbow lake gradually fills to become a marsh until it eventually dries out completely. Aerial photography usually shows a flood plain to be made up of loops of ridges and depressions, each marking

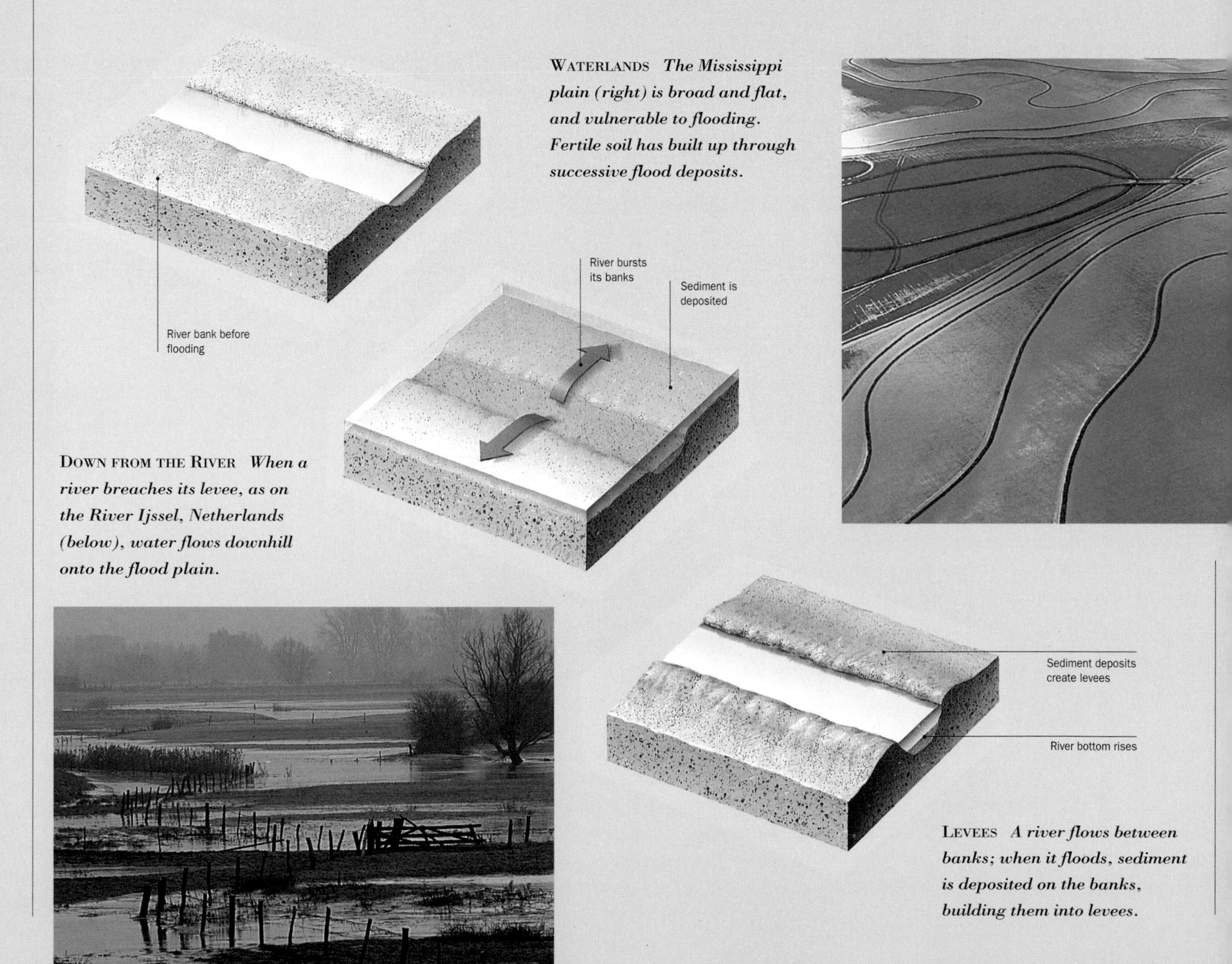

WATERLANDS *The Mississippi plain (right) is broad and flat, and vulnerable to flooding. Fertile soil has built up through successive flood deposits.*

DOWN FROM THE RIVER *When a river breaches its levee, as on the River Ijssel, Netherlands (below), water flows downhill onto the flood plain.*

LEVEES *A river flows between banks; when it floods, sediment is deposited on the banks, building them into levees.*

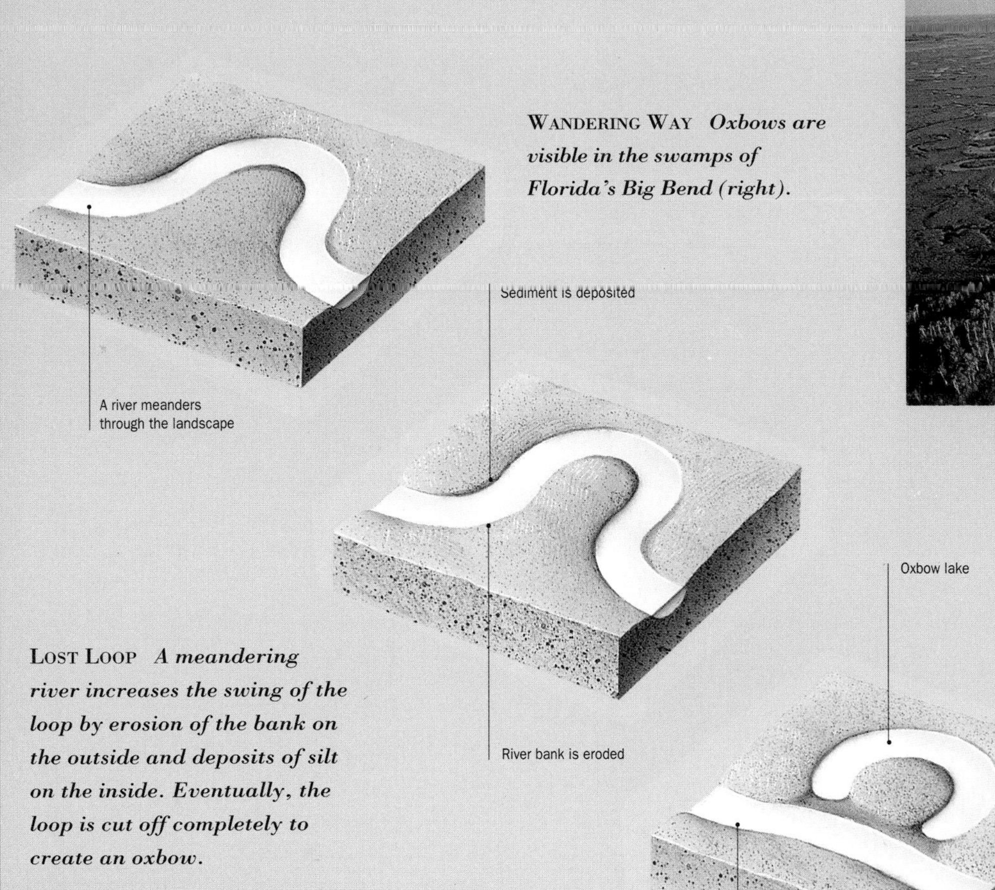

WANDERING WAY *Oxbows are visible in the swamps of Florida's Big Bend (right).*

LOST LOOP *A meandering river increases the swing of the loop by erosion of the bank on the outside and deposits of silt on the inside. Eventually, the loop is cut off completely to create an oxbow.*

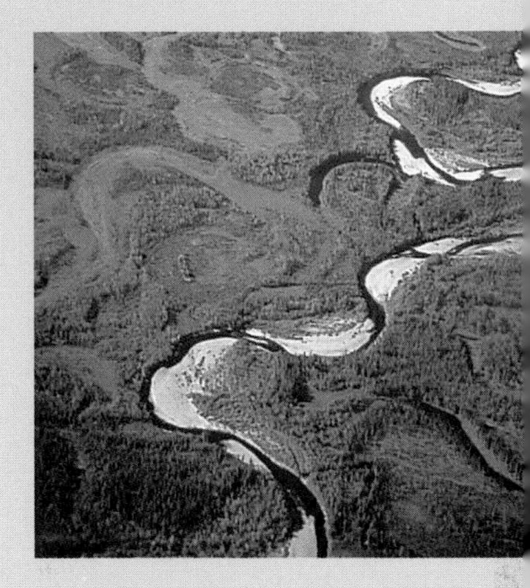

ALASKAN PAST *Traces of old oxbows show how this flood plain has built up.*

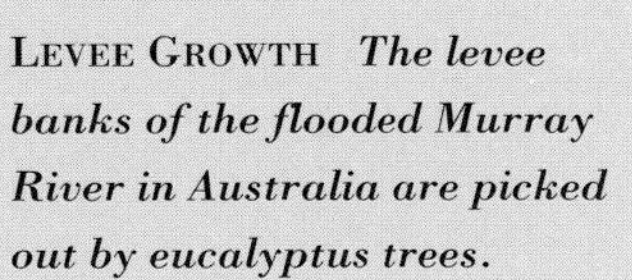

LEVEE GROWTH *The levee banks of the flooded Murray River in Australia are picked out by eucalyptus trees.*

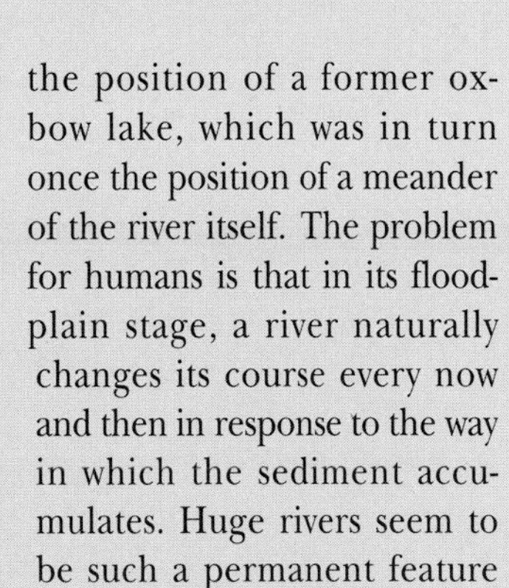

the position of a former oxbow lake, which was in turn once the position of a meander of the river itself. The problem for humans is that in its flood-plain stage, a river naturally changes its course every now and then in response to the way in which the sediment accumulates. Huge rivers seem to be such a permanent feature of their landscape that settlements and industries are built beside them on the assumption that nothing is going to change. When things do change, however naturally, this is regarded as a disaster if it has any effect on the people living there.

To make the Mississippi flood plain more amenable to settlement, there have been enormous engineering projects. The banks are clad in concrete slabs, called revetments, to prevent the erosion of the outsides of the meanders. These certainly work, as can be seen in the waters of the Mississippi, which are so much clearer and freer of material eroded from the banks than they used to be. The effect is particularly apparent where a stream that has not had such treatment joins the main river: the suspended mud of the stream can be seen as a swirling brown cloud where the two waters mix.

OXBOW *The still waters of an oxbow lake in New Zealand support thick wetland vegetation in the swampy soil.*

MOBILE HOMES *The floods of 1953 left caravans at Skegness in Lincolnshire washed up like empty boxes along a shoreline.*

water, was brought to a halt when the engine collided with a wooden bungalow from the park. Hundreds of people in the area were trapped on the roofs of their homes and clung there for days, while rescuers worked around the clock to reach them. Rowing boats in streets became the familiar image of the disaster. Rescue pilots could not recognise the area, since the traditional coastline had disappeared below the water.

Farther south, the River Thames forms a funnel-shaped estuary. The surge of water pushing up the estuary was squeezed between narrowing banks, with the result that its surface became higher and higher as it neared the city of London. The sea defences of Canvey Island were breached at about midnight and 58 perished there, as the entire island was submerged. Walls built along the banks of the Thames to keep flood water at bay were overtopped, and thousands of houses in the eastern part of the city were flooded. In the centre of the city the defences held, but only just. Had the flood defences been breached, the underground railway system would have been flooded and put out of action for months. The tunnels did have watertight doors, but these had been put in place during the Second World War as a defence against bomb blast. They were the wrong design, and opened the wrong way to hold back flood water from the river.

In East Anglia 24 500 people were evacuated, and 400 houses were completely destroyed. Some 530 sq miles (1350 sq km) were flooded, some for many months afterwards, and 307 people lost their lives. The salt water poisoned the farmland, leaving approximately 170 000 acres (68 800 ha) ruined. In some places, it was not even possible to graze cattle on the salt marshes for two-and-a-half years afterwards.

On the coast of the European continent, where the North Sea narrowed and the winds were blowing directly on shore, the results were even worse. Between Rotterdam and the estuary of the Scheldt, the sea broke through the dykes in 67 places simultaneously a few minutes before dawn, damaging more than 310 miles (500 km) of sea defences. The sea engulfed 133 houses in a few minutes and flooded an area of 625 sq miles (1600 km^2) – about 6 per cent of the arable land surface of the Netherlands.

DUTCH FLOODS *In 1953 the dykes (above) built to contain floods failed, and the Dutch polderlands were submerged.*

Stories are told of incredible heroism. A young couple, due to be married the following day, plugged a hole in a dyke with their own bodies; the woman died and her fiancé ultimately went mad. In the town of Kortgene, a hundred fishermen linked arms and pressed their backs to the failing dyke, but to no avail. In the Netherlands, 1800 people died and 50 000 head of cattle were lost. Some 72 000 people were evacuated and 43 000 homes destroyed. London had been spared, but the great industrial metropolis of Rotterdam was not so fortunate. These floods were the worst that the country had suffered for 500 years.

And they could happen again – since the conditions of a particularly high tide and a storm surge are not unique. If they did, the consequences could be far worse. Despite the fact that defences have been rebuilt and strengthened, and the Thames Barrier has been built at Woolwich, many of the defences were erected hurriedly after the 1953 disaster – hastily designed and shoddily built – and 40 years later, they are decaying.

The greater problem is that south-east England is slowly sinking, due to normal geological processes, and that the sea level is rising, possibly due to global warming. In 1990 the Intergovernmental Panel on Climate Change estimated that by 2030, the sea level will have risen by 7 in (18 cm), and that by 2070 it will have risen by 17 in (43 cm). Similar figures were produced independently by the US National Research Council in the late 1980s. It has also been

NATURE OR HUMANS *In Italy heavy rainfall bursts a dam that was not designed to take such pressure, causing extensive flooding downstream.*

estimated that, with the higher sea levels, a storm surge coupled with a high tide, causing a disaster like that of 1953, would now kill 3000 people.

DAM FAILURE

In addition to rain-swollen rivers and to the combined effect of winds and tides, the third major type of flood is caused by the burst of a dam. Since dams are man-made structures, the resulting disaster is usually classed as man-made rather than natural – unless, of course, the dam-burst is the result of a landslide into the lake.

Dam-bursts happen when the volume of water in the reservoir exceeds that for which the engineers planned, and so they study climatological records carefully in order to check that rainfall should not bring more water than can be handled by the system. It is not just the human observations and written records that are scrutinised. Recent geological deposits are an important source of information, too. Sediments deposited by floods are distinctive, and the heights to which these deposits build up are an indication of the depth of flood waters at some time in the past.

For example, floods can cause serious problems in Australia's arid hinterland. In January 1974 New South Wales and Queensland were struck by some of the worst floods during this century. Some 12 in (30 cm) of rain fell in 24 hours in areas that are normally completely dry. Dry river beds were running 20 ft (6 m) deep, and standing water lay around Alice Springs in central Australia – something almost unheard of in history. Studies of Australia's interior deserts have shown flood deposits that suggest that a 'superflood' caused by abnormal rainfall sweeps the continent every now and then. These deposits can now be dated, using radioactive analysis and carbon dating from engrained layers of soot from bush fires. Superfloods that were previously thought to have happened only once in several thousand years now seem to have taken place about once every 200 years, making them a far more dangerous threat than was hitherto appreciated. Such a superflood would be beyond the engineered limits, for

STREETS OF WATER *Communities built on low-lying lands are always vulnerable to flooding, as in Shepparton, Victoria, in 1993.*

example, of the Warragamba dam, which was built to supply Sydney's drinking water. Dam failure here would not only deprive the city of its water supply, but would threaten 56 000 houses built downstream from it.

Some dams are completely natural, and these can burst, too. Probably the greatest such disaster in North America took place between June 1971 and September 1973. The dates are imprecise because the area was uninhabited and no one witnessed the event. In the Coast Range of British Columbia, in Canada, 137 miles (220 km) north of Vancouver, lies Lake Klattasine, a glacial hollow filled by melting ice from a glacier. The lake was held back by a dam of moraine – rocks and debris left behind by the glacier – 65 ft (20 m) high. At some point the dam collapsed and 60 million cu ft (1.7 million m^3) of water drained from the lake into the downhill valleys. The wall of water transported a freight of rocky debris and washed out the valley slopes, causing landslides and adding to the debris that was carried along. When the torrent reached the valley of the Homathko River, it slowed down. Between

Australian Floods *Farms are inundated in the great floods of Victoria, 1993.*

70 and 140 million cu ft (2-4 million m³) of rocks and soil spread out and were deposited in a debris fan that was 65 ft (20 m) thick in some places.

The Human Experience

For the people caught in a flood, the experience is one that they will never forget. They see their property and possessions destroyed, and sometimes their relatives and friends killed. The authorities must react quickly, first to ascertain the extent of the damage and then to work out the best way of dealing with it. Since a flood tends to disrupt communication, it is often difficult to find out quickly just how big the problem is. A control centre and field hospital are set up as a matter of urgency. After medical aid has been supplied, food and shelter are the next priorities for the wet and hungry survivors.

The pumping out of water and drying of property is only one aspect of the follow-up repair operation. Debris has to be cleared from the affected area. The biggest problem, however, is likely to be mud. In its normal flow, the Huang He is so thick with mud brought down from the interior of the continent that it can form a slurry consisting of about 40 per cent solid material – which is what gives it its name, 'yellow river'. In Italy after the 1966 flood, an estimated 500 000 tons of mud were deposited in the streets and cellars of Florence. A river in flood carries much more suspended material than one flowing normally; when the flood spreads over the surrounding land, all this material is deposited as mud. The damage caused by the mud is usually much more expensive to put right than the damage caused by the water itself.

Preparation for these events is much better than repair afterwards, and being prepared involves understanding the workings of nature. Hillsides that have been cleared of forests have soils that are not broken up by tree roots. As a result, water flows quickly here and fills the local streams much more rapidly than under natural conditions. Furthermore, in some places the cleared areas have been drained by farmers to encourage the growth of grass for grazing animals. The underground drains they put in also carry away the rain water much more quickly than normal, filling the local rivers.

Precautions *Contour ploughing reduces run-off from farmland, and flood-relief channels in Los Angeles (bottom) anticipate heavy flooding.*

Roads and other waterproof surfaces, such as car parks, and even large expanses of roof, also lead to rapid run-off. And all these factors have to be taken into account when building in flood-prone areas. Some modern farming techniques involve 'contour ploughing', in which the ploughing of the soil takes place along the hill rather than up and down the slope. This tends to trap the downhill flow of water in the furrows and to slow its run-off to the rivers. Where a river is known to flood, the local peoples often build houses on stilts that hold them above any expected flood water surface.

Flood-relief channels, like those that Leonardo da Vinci proposed for Florence in 1500, are useful devices. Los Angeles makes great use of the wide, dry, concrete canals that have become as much a part of the characteristic Los Angeles cityscape as the Eiffel Tower is in Paris. 'It never rains in southern California', goes the song. But it does, especially in late winter and early spring. The mountainous and otherwise arid environs of Los Angeles are well poised to generate the typical flash floods of desert areas. Drinking water is stored in reservoirs in the mountains round about, and any flood water from them, or from the mountains, would be adequately taken care of by this system of flood-relief channels.

Where dams are situated upstream of major towns, the flow of water in and out of the reservoir must be monitored and controlled. One of the reasons for the Florence disaster of 1966 was the mismanagement of two major reservoirs on the Arno, upstream from the city. Engineers at the higher dam saw how deep the reservoir was becoming, but were too late in opening the sluices to let the excess run away. The result was a sudden influx into the lower reservoir, where engineers had to release all the water at once or watch their dam fail.

Unstoppable Power *Mountainous regions, such as Washington state, can be subject to torrential valley floods.*

Sea Cities *Many of the world's greatest cities, such as Bangkok and Venice, lie close to sea level and are prone to flooding.*

A Watery Future

As with all natural disasters, the onset of flooding is greatly aggravated by increasing populations. Bangkok, the capital of Thailand, is growing rapidly. In 1900 it covered just $\frac{1}{2}$ sq mile (1.3 km^2); by 1958 it had expanded to 37 sq miles (96 km^2), and today it exceeds 180 sq miles (460 km^2). Bangkok is built on the flood plain of the Chao Phraya River, about 15 miles (25 km) from the sea. Over the last two centuries, a system of drainage canals has been built to reduce the likelihood of flooding by the river. However, these drainage canals have proved to be totally inadequate.

To satisfy the needs of its 5.5 million inhabitants, Bangkok obtains water by wells sunk into the ground immediately beneath the city. The bedrock consists of layers of marine clays and sandy sediments and when these dry out, it shrinks. Last century, the water table lay only a few yards beneath the surface and extraction of water was easy. Today, however, it is becoming deeper and deeper, as 10 000 wells in the city each extract 46 cu ft (1.3 m^3) of water per day. The result is that the whole of the city is subsiding, in some places at a rate of about $5\frac{1}{2}$ in (14 cm) per year. It has been estimated that, if it continues at this rate, the entire city will be below sea level by the year 2000, leaving it vulnerable to floods from the sea, if the sea defences are breached, and from the river.

A similar disaster is threatening the northern Italian city of Venice. Overuse of ground water has caused the city to subside by about 5 in (12.5 cm) in the last 50 years. This has combined with a natural rise of $3\frac{1}{2}$ in (9 cm) in the level of the Adriatic Sea over the same period. Venice now floods whenever there is a high tide combined with heavy rain and a storm surge. Tokyo, likewise, is sinking at a rate of about 6 in (15 cm) per year because of withdrawal of ground water. The Japanese capital could be in much greater danger, however, because of its position in an earthquake-prone area and in a typhoon region.

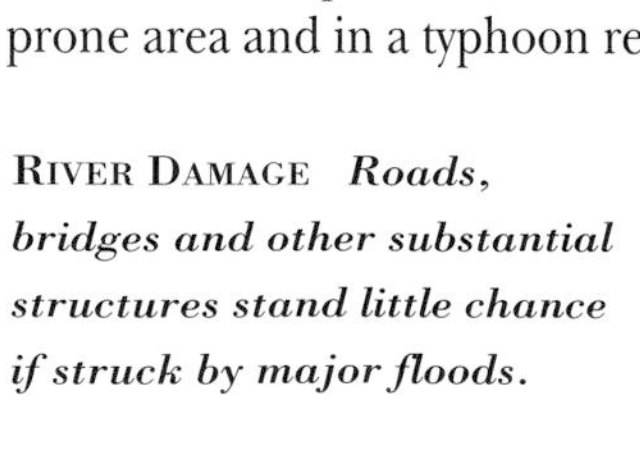

River Damage *Roads, bridges and other substantial structures stand little chance if struck by major floods.*

CLIMATE CHANGES AND DROUGHT

The human activities of a region – in particular the farming – are usually dependent on the climate. If this changes, the effects can be devastating: drought, crop failures and starvation can lead to human suffering on a massive scale.

The Sahara Desert was once a fertile plain. Rock paintings in the arid Tassili plateau in southern Algeria show pictures of antelope, giraffes and other grassland-living animals – eyewitness evidence of the local fauna some 8000 years ago. There are also scenes of herds of cattle being driven along, showing that where there is now nothing but hot, sun-baked rocks and dry sand, there was once grazing land.

Lake Chad, which now covers an area of just 970 sq miles (2500 km^2), spread over an area several times greater than that between 12 000 and 6000 years ago. Not far away, great trees grow in waterless gullies. These trees are thousands of years old and must have germinated when the soil was moist enough to support them. Over the centuries, the land became drier, and the water table dropped away. The trees' roots, however, continued to grow down as the water table fell, just keeping pace with it. Now the trees stand in desert aridity, nourished by roots that go way down deep to the distant water supply.

Satellite surveys have revealed the sinuous courses of ancient riverbeds below the sands of Arabia, rivalling the breadth of the mighty Nile of today, and flowing into the Persian Gulf in a delta that once covered the area of modern Kuwait. These waters stopped flowing sometime between 11 000 and 5000 years ago. However, even about 2000 years ago, the area that is now the Sahara desert was hunted for lions and ostriches to provide sport for the arenas of the Roman Empire. And the region to the west of the Nile was renowned as Rome's breadbasket. Nowadays things are very different. The Sahara is regarded as the epitome of dryness and sterility – and it is growing in area almost as we watch.

DRAMATIC CHANGE *Ancient rock paintings of grazing herds at Tassili suggest that the Sahara was once a fertile region. Now, sand dunes and city collide.*

The region on the desert's southern boundary is known as the Sahel and consists of semidesert and patchy grass, where the rainfall averages about 20 in (510 mm) per year. Chad, Niger, Mauritania, Mali, Burkina Faso and Gambia – the countries located in this area – are all poor and subsist on cattle rearing. The Sahara desert itself is spreading southwards into this region at a rate of 3 miles (4.8 km) per year, with the result that, over a period of 20 years, Mauritania has lost almost 75 per cent of its grazing land.

The expanding populations of these countries have caused an increase in cattle grazing and in the felling of trees for firewood, and both of these factors have encouraged the expansion of the desert. At the

PARCHED EARTH *When drought settles on the land, cracked, sun-baked soil, such as in this river, is typical.*

Frozen Solid *Frost fairs were a familiar sight on rivers such as the Thames in the mid 19th century.*

same time, many communities have adapted to growing cash crops, with the subsequent need for irrigation, rather than following their traditional, self-sustaining, nomadic way of life. However much these changes have exacerbated the situation, human influence has been less wide-ranging in its effects than a changing climate.

CLOUDING THE ISSUE

One of the many factors that make the prediction of climatic change such an imprecise science is the effect of clouds. As temperatures rise, more water may be evaporated from the oceans to form clouds. What effect this increased cloud cover would have is open to debate. On a summer's day, clouds reflect away the incoming solar radiation, thereby keeping ground temperatures down. On a winter's night, however, they act like a blanket, trapping what heat is already contained in the ground and preventing the formation of frost. Both these effects – the reflective effect and the insulation effect – will come into play in the future, but climatologists cannot predict what the balance will be.

A Warming World

In the mid 19th century, scientists began to keep records of the worldwide climate. But even before then, when climate observations were less coordinated, it is evident that there had been several periods of great cold in historical times, in which rivers such as the Thames (which has not frozen solid since the winter of 1814) were so thick with ice that fairs could be held on them. One such cold period happened in the 17th century, and allusions are made to it in Shakespeare's plays where winter is described in dramatic imagery indicating conditions far harsher than anything we experience now. Over the last century, the world has warmed up by about 0.5°C (0.9°F). Eight of the world's warmest years this century have occurred since 1983.

In the long term, the Earth's climate changes constantly. The Ice Age, which gripped the Earth first about 1.7 million years ago, was not an uninterrupted period of cold. It embraced about 20 fluctuations, in which the coldest times – called the 'glacials' – were separated by interglacials lasting up to tens of thousands of years when the climate was warmer than it is at the moment. During these warm spells, hippopotamuses wallowed in the River Thames, where London now stands. The end of the last glacial interval, about 10 000 years ago, is regarded as marking the end of the Ice Age; since then climates have been becoming warmer. During these 10 000 years, the face of the Earth has changed completely. The sea level is about 330 ft (100 m) higher than it was then, habitats have changed from tundra to forest, and species have evolved or become extinct – among the mammals, at least 32 genera have died out. However, the journey out of

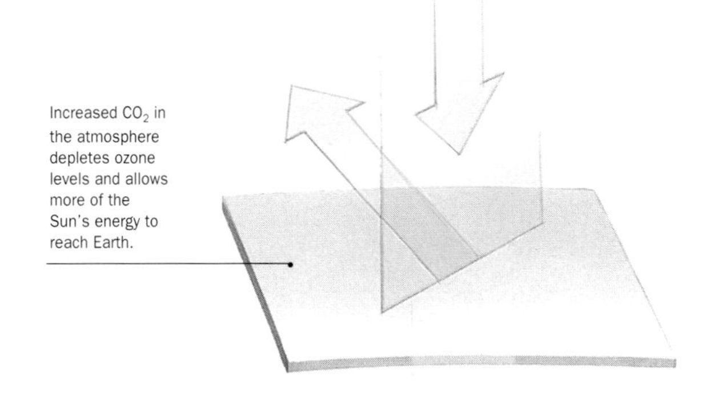

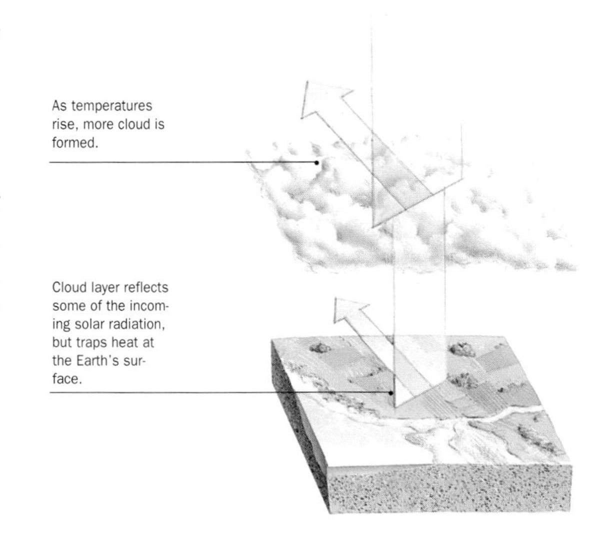

Warming up *Too much sunlight passing through the atmosphere upsets the balance between heat absorption and heat loss at the Earth's surface.*

PEOPLE OR NATURE TO BLAME

Desertification, the term given to the increase in desert areas over the years, usually carries an implication that it is humans who are to blame. Recent research, however, by the National Oceanic and Atmospheric Administration in Boulder, Colorado, suggests that dryness in the Sahara runs in cycles of several decades. A long-term drying out such as we have witnessed recently could, therefore, be a perfectly natural state of affairs. The history of the peoples of such places shows that in the past they have been able to survive by adapting to the natural rhythms.

SURVIVAL *A nomad draws water from a local well in Mali.*

Today the populations are so much greater, and this inevitably increases the pressure on the environment. It may be necessary in the future for any aid to be based on an adaptation to these natural rhythms. Rather than drilling wells in urban areas or introducing crops that can flourish only with the back-up of high technology and irrigation systems, it may be better to try to find out how previous desert civilisations survived and develop their strategies on a scale appropriate to modern populations.

the Ice Age has been very unsteady and has often been broken by particularly cold spells. Indeed, in the 1970s it was believed that the Ice Age was returning and that we were merely living in an interglacial. It was argued that, in another glacial stage, the glaciers would return to cover the northern continents – a natural disaster of vast proportions.

Today, however, the pattern of climate change is becoming clearer. The same meteorological institutions that predicted a continuation of the Ice Age now warn of a global warming. In the subtropics, from the West Indies, through north Africa, India and the Far East, there is much less rain than there used to be, while there is much more rain falling farther north. The situation is mirrored in the equivalent latitudes of the Southern Hemisphere, with the wetter areas moving farther south. The rising temperatures are having other dramatic effects. Mountain glaciers and the polar ice caps are melting. Without the brightness of a permanent covering of snow and ice, which reflects a great deal of the sun's heat, temperatures around the globe would rise more quickly. The melting ice is raising sea levels and meltwater from the Arctic may also divert the Gulf Stream, a warm water current across the Atlantic which keeps north-west Europe several degress warmer in winter than equivalent latitudes in Canada. Without the Gulf Stream, places like Britain would become locked in ice during the winter.

THE EFFECT OF PEOPLE

Natural climate change usually takes place over long periods of time. However, in the shorter term, it seems that human influence is having a much greater effect. Factories, car exhausts and power stations all emit gases into the atmosphere. Chief amongst these are carbon dioxide, water vapour, methane and nitrous oxide, which warm the surface of the Earth by allowing the Sun's rays to pass through the atmosphere and then trapping any heat that is usually radiated away – in a process known as the 'greenhouse effect'. As a result, temperatures are increasing much more quickly than they were.

The conditions of the 'Little Ice Age', which occurred from the mid 17th to the late 19th centuries, were caused by temperatures that were, on average, only 1°C (1.8°F) lower in northern Europe than today, but since 1900 the average world temperature has increased by 0.5°C (0.9°F). An international panel of scientists has predicted that, by the year 2100, the Earth will be 5°C (9°F) warmer than it is now. The fact that these are average temperature rises implies an increased frequency of extreme conditions. For example, an average increase of just 2°C (3.6°F) would double the chances of heatwaves in the major cities of the United States. Such a fast rise in average temperature could wipe out whole ecosystems and any animals unable to evolve fast enough to accommodate the change in conditions. It would also lead to unstable weather conditions. If the temperature rises before the ice caps melt, there will be a great temperature difference between the hot areas of the Earth and the cold, which will in turn increase the frequency of tornadoes, hurricanes and the occasional 'superfloods', caused by abnormally high rainfall in certain areas.

The major 'greenhouse' gas is carbon dioxide, generated by the combustion of carbon in the form of fossil fuels – coal, oil

ATMOSPHERIC CHANGE *A haze of pollution hangs over Sao Paulo, Brazil. Far right: A copper smelter discharges sulphur dioxide into the atmosphere.*

and gas. These fuels represent natural carbon dioxide that was extracted from the atmosphere and locked up in the Earth by biological and geological processes that took many millions of years. Now, in a mere 200 years since the Industrial Revolution, vast quantities have been returned to the atmosphere, and it is difficult to see what can be done about the problem. Many scientists suggest that it will stimulate the growth of more vegetation, thereby locking it up again.

However, this new vegetation will in turn die and decay, and return the carbon dioxide to the atmosphere once more. Only if all this vegetation was buried and transformed into coal and oil would the process do any good. Another possibility is that the oceans will represent a sink for excess carbon dioxide. The gas will become dissolved in the ocean water and stimulate the growth of microorganisms that will eventually convert it into the mineral calcite, thereby locking it up in layers of limy mud on the ocean floor. The great thicknesses of limestone found today testify to the fact that this process has already happened in the geological past.

The Second Horseman

Among the Four Horsemen of the Apocalypse, Drought and Famine go hand in skeletal hand. They may be the result of natural conditions, but it is usually the human input that makes them so much worse than they might have been. Perhaps the greatest famine of modern times was that which struck Sudan and Ethiopia between 1983 and 1988, during which an estimated 1 million people died. However, its causes go back decades. After a drought in the Sahel between 1968 and 1973, the United Nations and other aid agencies began a programme of drilling water wells close to towns, which quadrupled the population of the urban areas. Most of the new settlers had previously been nomadic peoples who could no longer continue their traditional way of life. Instead, they had to settle and farm in areas that were unsuitable.

The problem was compounded by the oil crisis of 1973 and 1974, which crippled the economies of the industrialised Western world and prevented them from sending the aid that would normally have been forthcoming to relieve the famine in Africa. Several hundred thousand people died. In 1980, no rain fell at all in Ethiopia, the rivers dried up, and an estimated 50 per cent of the country's cattle starved to death.

Refugees *Cattle starve in Sudan, and drought in Ethiopia (bottom) creates large numbers of refugees.*

Dust Storm *In drought conditions the soil just blows away as dust.*

When the rain did fall in 1984, the aid agencies relaxed but, by this time, the ground was so badly ruined that the rain simply washed the soil away.

Dust storms are one effect of drought. Known as a haboob in North Africa, this phenomenon occurs when high winds, possibly generated by storms, sweep up the loose dust and sand from the top of dry soil and swirl it about. The storm has a rolling front and can be 300-400 miles (480-640 km) in diameter, and may carry dust to heights of 8000-14 000 ft (2400-4270 m). It can shift 100 million tons of sediment in a few hours, piling up drifts of sand 12-15 ft (3.7-4.6 m) thick in places. Dust from the Sahara has even been known to fall in South America. In the thick of such a storm, visibility drops to 1/4 mile (about 400 m), but it can become much worse. The air can be so

Creeping Desert *Houses at the edge of the Sahara are periodically choked by sand and need clearing out.*

thick with dust that it suffocates any living things. For example, a recorded dust storm in Colorado in 1895 choked 20 per cent of the state's cattle to death. Even after the winds have stopped, the dust can remain suspended in the atmosphere for days. Wind storms of this type can erode 20 million tons of soil each year from the farmlands of the United States, and annually render useless 1 million acres (400 000 ha) of Russia's agricultural land.

The sequence of events known as El Niño, 'the child', because when it occurs it generally happens at Christmas, is another form of climatic change with disastrous consequences. Usually, the tropical belt of the Pacific Ocean is warmer around the Philippines and cooler around the coast of South America because the Humboldt Current brings cold waters northwards from the deep, Antarctic seas. Occasionally, however, this pattern reverses for a limited time. The warmer waters create dry conditions in Africa, southern Asia and Australia.

The droughts of Indonesia in 1983 and the 'Ash Wednesday' fires of Australia in the same year were a direct result, as were the rainstorms in California and the drought and bush fires of the Galapagos Islands in the late 1980s that had such a severe effect on the wildlife of this archipelago. El Niño's impact is felt much farther afield as well, the great droughts in Africa in the 1980s coinciding with the onset of El Niño. Whereas the phenomenon used to be expected once a decade or so, it is now much more frequent.

Other Authorities – Other Views

Some of the data on climatic change is open to interpretation, and there are still climatologists who refute the idea of a greenhouse effect. Their arguments are *(continued on page 134)*

On the Move ***Masai people take their cattle and goats in search of new pastures.***

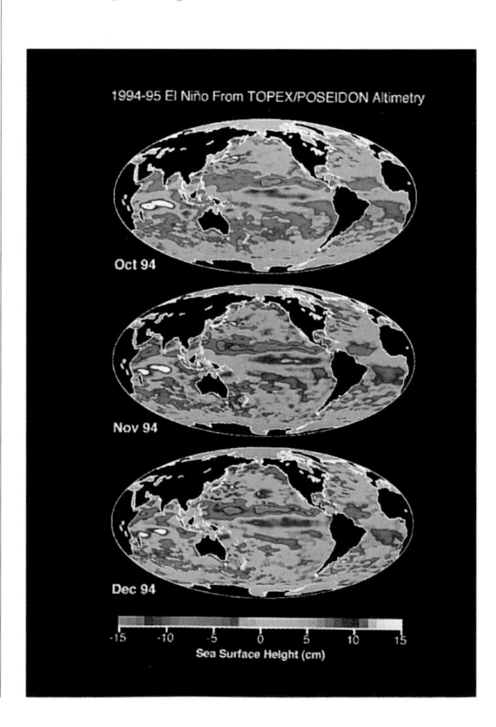

El Niño ***The red tongue in the Pacific shows the development of the 1994 El Niño.***

THE TETHYS LEGACY

Abandoned Hulk ***A ship is left high and dry as the Aral Sea evaporates, starved of incoming waters by irrigation schemes upstream.***

The Mediterranean is the largest of a series of inland seas, including the Black, the Caspian and the Aral Seas, that stretch eastwards from the Atlantic Ocean. Geologically, they are all that remain of the mighty Tethys Ocean that once reached between the continents of Europe and Asia to the north, and Africa and India to the south, and was squeezed out of existence about 50 million years ago as the continental masses collided. As well as providing natural reservoirs in a desert landscape, the expanses of water provide a tempering effect on local climates.

The Aral Sea has been shrinking for 30 years. What was once the fourth biggest lake in the world, with an area of 26 390 sq miles (68 320 km^2), has, since the 1960s, lost 70 per cent of its water through evaporation, and in some places its shorelines are now 50 miles (80 km) away from the water. The fishing fleet is landlocked on the dry mud flats; and in any case, the fish stocks have died. This has happened because, in the 1960s, the two rivers that flow into the Aral Sea were diverted to irrigate the cotton and rice crops of Kazakhstan, Uzbekistan and Turkmenistan; the irrigation ditches were so badly made, however, that they waterlogged the agricultural land.

The Culprit ***Irrigation for the intensive production of cotton in Uzbekistan led to the loss of water from the Asian inland seas.***

The ameliorating effect on the climate has disappeared, too – summers are hotter and drier, and winters colder and longer. There are now about 150 days without rain each year, as opposed to only about 30 previously. Dust storms are frequent, disturbing about 150 million tons of dust and salt per year from the dry seabed. About half of this is dumped on nearby agricultural land, while the rest drifts away in the atmosphere. As well as the dust and salt, pesticides and fertilisers that had drained into the sea are now being blown about, further poisoning the land. The full scope of the disaster was not known to the outside world until the days of Glasnost, when the irrigation scheme was halted in 1986. Salt-resistant plants are now being grown to stabilise the dry seabed, and canals are being dug to drain off the waterlogged agricultural land into the Aral Sea.

Meanwhile, 310 miles (500 km) to the west, the Caspian Sea was facing a similar disaster. The surface of the Caspian fluctuates naturally, rising and falling every 70 years or so. In 1976 it had fallen particularly low, to about 295 ft (90 m) below mean (average) sea level, and the Soviet government panicked, fearing a disaster similar to the one that had struck the Aral Sea. Against the advice of hydrologists, a grandiose scheme to remedy this was begun in the late 1970s.

Salt Flats ***Thick deposits of salt are left as the water evaporates from Kara-Bogaz, an inlet of the landlocked Caspian Sea.***

One particular area of the Caspian – the Kara-Bogaz – is a vast shallow gulf into which water has always flowed and then evaporated away, leaving deposits of salt that have been a mainstay of the local chemical industries. It was thought that if the gulf could be walled off and allowed to dry out completely, then the water from tho Caspian that would otherwise have evaporated away here would be saved. A vast dyke across the entrance was completed in 1980. But within two years, this was shown to have been a mistake. Salt and chemicals from the dried-out gulf blew up to 125 miles (200 km) inland, poisoning crops. And the underground brines that were so important to the chemical industries ceased to flow.

Then, as the hydrologists had predicted, the level of the Caspian Sea began to rise again as part of its natural cycle. Without the Kara-Bogaz to aid the evaporation of the water, however, the level rose higher than usual, flooding fishing villages and oil wells around the coast, and cutting the main railway link. A state of emergency was declared in 1991 and the dam demolished – but not before the natural systems of the sea had been permanently disrupted.

SEASHORE DROUGHT *The skeleton of a marine iguana that was a victim of starvation on the Galapagos Islands during the El Niño of 1982.*

based on the premise that most climate observations are obtained from stations sited in built-up areas, which record higher temperatures because urban areas are warmer than rural regions. They argue that as urbanisation has increased over the past century, so local recordings in newly built-up areas have become distorted.

There are even scientists who suggest that the greenhouse effect might be no bad thing. They point to the fact that higher temperatures existed in the Earth's geological past and that these coincided with particularly flourishing times for animal and plant life. The world temperature predicted for the end of the 20th century would be similar to that of a warm period 6000 years ago at the end of the Ice Age, when increased rainfall made the Sahara fertile. The temperature in 2050 would be the same as that in the Pliocene Epoch, about 5 million years ago, which featured the last fine conditions before the onset of the Ice Age. This would bring about wetter as well as warmer conditions, and allow grain crops to be produced in areas such as Siberia or Canada that are currently too cold. As more water is evaporated from the oceans, more will be precipitated as rain, increasing the annual rainfall in the Sahara by something like 12 in (30 cm), and stimulating the growth of plant life there.

These optimistic scenarios are very much minority views. The consensus is that global warming will produce famines and a massive refugee problem, as low-lying countries such as Bangladesh are flooded by rising sea levels. Energy conservation and the reduction in the use of fossil fuels, together with the development of farming techniques that take the workings of the natural ecosystems into account, appear to be the way forward into the future if such disasters are to be avoided.

OIL MONEY *Wealthy desert countries such as Saudi Arabia can afford complex irrigation schemes.*

4

FREAKS OF NATURE

WARNING FROM ABOVE *Comets were traditionally regarded as harbingers of doom.*

THE TRADITIONAL ELEMENTS OF THE ALCHEMIST WERE EARTH, AIR, WATER AND FIRE. EACH HAS A DESTRUCTIVE POWER THAT, IF NOT BROUGHT UNDER CONTROL, CAN LEAD TO TERRIBLE NATURAL DISASTERS. OF THE FOUR, PROBABLY THE MOST FRIGHTENING IS FIRE. THE TERROR OF FIRE STEMS FROM THE FACT THAT IT IS A VERY FAMILIAR PHENOMENON. ITS EFFECTS ARE WELL KNOWN, AND SO THE PAIN AND DAMAGE CAN BE ANTICIPATED. HOWEVER, THERE IS ANOTHER THREAT TO OURSELVES AND TO OUR CIVILISATION. THIS ONE IS TERRIFYING BECAUSE ITS RESULTS ARE UNKNOWN. THE THREAT FROM OUTER SPACE CANNOT BE QUANTIFIED BECAUSE NO ONE HAS OBSERVED SUCH AN EVENT — AND LIVED.

ESCAPE *Wildlife flees as grasslands burn.*

WILD FIRE

Learning how to handle fire was one of the most important breakthroughs made by prehistoric man. But when a combustible material combines with oxygen, it can turn into an inferno within minutes, causing destruction on a grand scale.

Since the earliest times fire has been used for cooking, to make meat tender and more digestible; for warmth, to keep away the chills of the Ice Age winter; and for defence, exploiting the natural fear in which the animal world holds the naked flame. Even the less controlled use of fire was probably employed by our earliest ancestors. There is evidence that *Homo erectus* tribes in Spain 800 000 years ago used grass fires to stampede elephants into bogs, making them easier to kill. And the simplest type of agriculture is based on a slash-and-burn system, in which the natural forest is cleared by fire of its plants in order to make room for the crops. In modern times, non-industrial societies still make very creative use of fire. The bushmen of southern Africa burn grasslands so that when the grass regenerates as succulent green shoots, it attracts game animals into their area for them to hunt; while the Aborigines of Australia burn the bush periodically to make it safer – leaving fewer hiding places for snakes and other poisonous creatures.

FIRE AS ENEMY

Fire can be a good servant, but as the old saying goes, it is also a bad master. To most 20th-century people, it is something to be feared. Certain parts of the world are more prone to fires that rage out of control than others, with areas that have what is known as a 'Mediterranean climate' being particularly vulnerable. These areas include California and the countries bordering the Mediterranean Sea in the Northern Hemisphere and parts of Chile, South Africa and the southern parts of Australia in the Southern Hemisphere. All these regions lie on the coasts of continents, where westerly winds bring warm, moist air from the sea in winter, and summers are hot and dry. Winter conditions encourage the prolific growth of vegetation that dries out under the heat of the summer sun – producing potential fuel and then the conditions to make it inflammable. As a result, the countries around the Mediterranean Sea account for 90 per cent of Europe's wild fires. The warm climate, coupled with predictable weather patterns, makes such areas attractive for settlement, especially among people who do not have to work the land.

BUSH FIRES

Around Melbourne in February 1983, at the height of the Australian summer, there was a heatwave brought on by El Niño – a phenomenon caused by a temporary rise in temperature across the Pacific Ocean. The subsequent disruption of the balance between warm and cool ocean currents had

FRIEND OR FOE *Fire can be benevolent. Here a controlled fire clears the ground for new pasture in South Africa.*

DEATH OF A TREE *A bush fire engulfs palm trees as it rages beyond any control in a dry, tropical area.*

created unusually dry conditions in Australia. These lasted for many months and the vegetation, which consisted of grasses and eucalyptus woodland, dried out. On Wednesday, February 16, the temperatures rose to 40°C (104°F) and 180 fires broke out in the states of Victoria and South Australia. Anything could have started them – sunlight concentrated through a broken bottle, a discarded cigarette end, a spark from an electrical appliance. The grass was as dry as it could possibly be, and once the fires had started, they spread immediately. With the wind blowing at 40-50 mph (64-80 km/h) from the arid interior of the continent, the heat blasted through the grass.

Common Event *A bush fire sweeps through Kakadu National Park in Australia, an area well used to such an occurrence.*

Most of the 180 fires were brought under control, but ten of them could not be contained. Eucalyptus bark and leaves are very oily and ignite readily under these conditions. As the fire swept up the trunks of the trees, the burning twigs and leaves were wrenched off and swirled away, landing far off and igniting new areas of grass and trees. A blazing fire sweeps the air violently upwards, dragging in air and fuel from round about: one such firestorm was seen blasting up to heights of 1250 ft (380 m). Firefighters could not work in conditions such as this. No sooner had they put out one patch of fire, than another sprang up some distance away.

All around Melbourne the fire raged, bearing down on the city and engulfing the outlying communities. The little town of Macedon was built of timber, with little in the way of solid masonry, and the roofs were made of corrugated iron. Houses just became more fuel for the wild fire, and disintegrated in flames while the inhabitants fled. The fire generated by the burning walls was channelled through the buildings by the iron roofs and nothing could stop it. Some 1.24 million acres (500 000 ha) of land were destroyed and 76 people lost their lives in this disaster, which became known as 'Ash Wednesday'.

The Ash Wednesday fire, as with all wild fires in populous areas, was aggravated by the fact that the communities regarded fire as an enemy, to be eliminated at all times, rather than as a natural phenomenon of the region that could have been accommodated. Under normal conditions in the wild, plant debris on the ground is regularly consumed by natural fires. In built-up areas, however, fire has always been discouraged and, as a result, layers of fuel accumulate over decades. In Australia, when the appropriate conditions arose and the fires started, they were therefore far more intense and difficult to control than they would naturally have been.

Natural Processes versus Preservation

The way in which natural fires are dealt with forms part of the wider debate between 'preservation' and 'conservation'. The aim of preservation is to keep conditions exactly as they are. That of conservation, on the other hand, is to follow the natural processes as closely as possible, even if this involves allowing natural fires to change the environment from time to time. In one firestorm near Los Angeles in 1994, the temperatures soared up to 1400°C (2552°F), which was hot enough to set fire to the asphalt on the Pacific Coast Highway. One of the reasons why this fire had gone so completely out of control, stopping only when it reached the Pacific Ocean, was that it swept through Californian suburbs where small-scale fires had never been allowed to

Homes Threatened *A forest fire spreads through an area of scrubland and housing near Los Angeles.*

start so a lot of plant debris had collected at ground level.

Under natural conditions, landscapes that are prone to periodic burning contain plants that have adapted to this. Although their bark and leaves are highly inflammable,

IMMUNITY *Dryland plants like eucalyptus (right) and banksia (far right) are able to regenerate after a bush fire.*

the eucalyptus trees of Australia have a wood that is almost completely incombustible – just try burning eucalyptus logs on a garden bonfire. Such plants are accustomed to regular conflagrations. Indeed, some eucalyptus species will reproduce only if fire has weakened their seedpods and allowed the seeds to be shed. Furthermore, the chemicals in wood smoke, particularly ethylene and ammonia, stimulate the tree seeds into germinating: horticulturalists even manufacture fertilising compounds from them, marketing them as 'fireless smoke' to aid the germination of trees adapted to such environments. After a fire, many eucalyptus trees send up fresh shoots from their roots through the charred soil, thus ensuring that they have a good start on the process of recolonisation, before rivals have had a chance to move in.

FOREST FIRE

Coniferous forests, such as those that form the world's largest unbroken belt of forest – the backwoods of Canada and the taiga of Siberia – are also prone to forest fire. Under normal conditions, a natural fire tends to start high up on a hill, where trees are more exposed to lightning strikes or even volcanic activity, and burn slowly downhill over a relatively small area. A fire started by a bolt of lightning usually affects

an area of less than 1/4 acre (1/8 ha). Nowadays, however, most fires tend to start in lowland areas, where the forest is more accessible to tourists or forest workers. Human carelessness is now the most common cause of widespread fires in the civilised world, accounting for an estimated 90 per cent of all wild fires in the United States and over 4 million acres (1.6 million ha) of forest damaged each year.

Probably the most deadly forest fire ever took place in Peshtigo, Wisconsin. The summer and autumn of 1871 had been hot and dry, with 14 weeks of drought in the area. On October 8, conditions became so unstable that fires broke out all over the Midwest. A great fire in Chicago that night killed 250 people, but the most serious loss of life occurred in the settlements of the forested areas. Firebreaks had been cleared around these settlements, but they could not hold back the conflagration of countless fires. At least 1500 perished that night, and 4.25 million acres (1.7 million ha) of forest were destroyed.

CHICAGO FIRE *A drought that struck the whole of central North America caused the 1871 conflagration.*

In the aftermath of this tragedy, the National Forest System was established in an attempt to control such events. In 1872, Yellowstone was established as the world's first national park, with a system of management designed to ensure the natural identity of the area. However, it was not until

YELLOWSTONE FIRE *The disastrous fire of 1988 was exacerbated by human interference with the natural system.*

1963 that the policy of subduing forest fires wherever they happened was recognised as being contrary to the natural scheme of things. The great forests, including Yellowstone, were made up of a mosaic of different habitats representing areas that had been burned at different times in recent history and had then grown back again. When fire destroys an area, the vegetation grows back in a particular sequence, not reaching its original pre-fire state for about 300 years – with the result that the patches of a typical forest are all at different stages of regeneration.

The no-fire policy of Yellowstone was seen to have interrupted this natural pattern for almost a century. Since it is the

FIRE CASUALTIES

Wild fires are frequent, but individual fires tend not to kill large numbers of people. In the United States and Australia, it is estimated that fires of all types kill about one person in 25 000, most of whom die in building fires through suffocation and the inhalation of smoke and poisonous gases.

oldest trees that are the most inflammable, the whole forest was becoming increasingly fire-prone as the trees were allowed to become older. In 1972 the park officials instigated a programme of controlled burning. Whenever lightning started a fire it was allowed to burn, only being brought under control when settlements and human property were threatened.

This new policy was hailed as a success until the summer of 1988, when things went seriously wrong. Winter had been very dry, but the wet summer that usually follows a dry winter there failed to materialise. In fact, it was the driest summer since the dustbowl days of the 1930s, with temperatures reaching 32°C (90°F) and high winds, some more than 100 mph (160 km/h), racing through the forest. When the fires began, the forest authorities watched them spread over greater areas than they had ever seen, and by July 15, more than 17 000 acres (6900 ha) had been consumed. Officials realised that the fires were not following the usual sequence of events and decided that day to fight every fire that developed. The dry conditions, and the leaf-litter that had accumulated over a century, made this an impossible job for the 9000 firefighters who had been drafted in.

The biggest fires are self-sustaining phenomena. The rising temperatures generated by the burning process produce strong convection updraughts that suck in oxygen and debris from the surrounding area. Loose branches, twigs and conifer needles are pulled into the blaze by these strong winds: at their most powerful, 'firestorms' have been known to pull over fire engines and suck firemen into the blaze. During August, experts in a variety of fields – meteorologists, botanists, geographers, computer programmers – were flown in from all over the country, and they worked with computer models to estimate the damage that would be done and what to do about it.

In fact, the blaze defeated all their predictions, since that summer's weather patterns were quite unlike any that had been seen in the area for a century. Their models greatly underestimated the area of forest that was likely to burn. In some areas, the fire advanced in uneven fronts at a rate of 10-12 miles (16-19 km) each day. Burning twigs and sparks, blasted upwards by firestorms, drifted for miles before settling and igniting spot fires in all kinds of unexpected places. By the end of the summer, more than 1.5 million acres (610 000 ha) had been consumed by fire, and it was not until the snows came in September that the fires began to subside.

Foresters and firefighters recognise three different types of forest fire: ground fire, surface fire and crown fire. In a ground fire, the organic component of the soil itself burns. This may consist of humus – the partly decayed plant material that is incorporated into the soil – or the living and dead roots and underground stems of the plants themselves. Since soil is an insulating medium that retains heat, a ground fire can be difficult to subdue once it has started. Surface fires are more familiar, with the loose debris of the forest floor, such as broken twigs and fallen leaves and needles, along with the growing grasses, ferns and other undergrowth, providing the fuel. In a crown fire, it is the tops of the trees that burn. Such fires can be difficult to control, since the seat of the fire may be too high to be reached by hoses, while the fire sweeping around the trunks and branches carries sparks and burning fragments away from the source to start fires elsewhere.

FIRE SURVIVORS

A natural fire can have a beneficial effect on a forest. For example, diseases and parasites do not flourish in fire-prone areas; and mistletoe, a serious pest of growing trees, cannot gain a proper hold if fires are frequent. On the other hand, hunting wasps, which prey on pests such as wood-boring and tree-killing beetles, make their nests in freshly burned wood. Without fires, these

THREE LEVELS *A fire burns on the soil surface (below); fire spreads up through the forest, (below right) and burns in the crowns of the trees (far right).*

NATURAL PROTECTION *The redwood's bark and cones are adapted to survive fire.*

wasps would decline and the beetles would be free to rampage.

Some forest trees, such as the redwoods of California, are adapted to survive fire, as are the eucalyptus of the Australian bush. A redwood tree has a spongy bark that can be up to 6 in (15 cm) thick, which insulates the wood from the fire so that it is not destroyed. If the outer surface of the bark is scorched by the searing heat of a forest fire, the living wood inside is protected. The coastal redwoods can regrow from fireproof burls – woody parts of the tree that sprout new growth. Furthermore, as with the eucalyptus, the redwood's cones open up shortly after a fire has passed and shed the seeds over the freshly cleared soil. Many of the redwoods in California's national parks have survived with these strategies for over 2000 years of frequent forest fires.

GRASSLAND FIRES

The great grasslands of the world stretch across the continents between the equatorial rain-forest belt and the desert belts of the tropics. Conditions here change, with the seasons, between the climatic extremes of the neighbouring areas; the wet season heralds a period of prolific growth as in the equatorial forest, and the dry season brings the searing desert sun. As with the Mediterranean-type climate, these conditions produce fuel at one time of the year and the environment in which it can burn at another.

The arid grasslands of the world – the savannah of Africa, the prairies of North America, the pampas of South America, the steppes of Asia – consist of grasses that are unharmed even if the leaves are destroyed. This is because the growing part of the plant lies underground, and if the leaves are grazed by animals or burned off by fires, then they can easily regenerate later. Grass seeds are spindle-shaped and slide

THERAPEUTIC FIRE *Grasslands are kept healthy by controlled burning.*

HOW NATURE COMPOUNDS A MAN-MADE DISASTER

When Reactor No. 4 in the nuclear power station at Chernobyl caught fire in 1986, the surrounding area was evacuated and the contaminated farmland, which had been managed and cultivated for centuries, was returned to nature. In its wild state, inflammable plant materials accumulated on the ruined earth, causing the outbreak of fires in the hot summer of 1992. About 136 fires raged across the deserted farmland, churning up the radioactive material that had polluted the landscape and carrying it high into the air on hot updraughts and smoke particles. As a result, it spread even farther than before.

POISONOUS RUIN *Reactor No. 4, now in ruins, produced contamination that was later spread by natural fires.*

into the ground easily. Some have long spikes that catch in the wind and spin them around, drilling them into the soil, away from any damage likely to be caused by grassland fire. Since dead plant material decays very slowly in such dry conditions, the nutrients are locked up and inert. Fire, however, breaks down the decayed material, liberating the nutrients as ash, which is absorbed into the soil much more readily, enriching it for the next generation of growth.

Grassland animals have their strategies for escaping fire, too. Most big, grass-eating animals are long-legged runners and can race away. Smaller grassland animals, on the other hand, tend to dig a burrow well below the fire itself in order to escape the conflagration.

FIGHTING FIRE

Fire needs three things – oxygen, heat and fuel – which are known to firefighters as the 'triangle of fire'. Take away one of these items and the fire is extinguished.

Fire-retardant foam, consisting of bubbles of nonflammable gas such as carbon dioxide, removes oxygen from the triangle. It tends to be used on man-made disasters such as oil spills, rather than on the much more widespread fires that are found in the open country. On heaths and grasslands, large teams of beaters are often used to beat down the flames with brooms and paddle-shaped tools, which smother the fire. It is important, however, to keep the teams in the area after the fire has been put out, to ensure that any residual heat does not re-ignite the unburned grass.

A more usual way of dealing with fire is to remove the heat from the triangle, normally done by dousing the fire with water. In a wild fire, this poses severe logistical problems. A landscape that is dry to start with may not contain enough water to be effective; and water might have to be pumped several miles from the nearest river. Tanker trucks can bring water to the site of the fire but only if there are adequate roads.

An effective but expensive alternative is to equip seaplanes and helicopters with dump tanks that can be filled with sea water, flown over the fire, and the water released. For aerodynamic reasons, only small quantities of water can be carried at a time, and so the use of this method needs to be carefully planned.

Firefighting by removing the fuel from the triangle of fire usually involves isolating the fire in as small an area as possible and letting it burn itself out. Firefighting teams clear firebreaks in advance of the front of

OUT OF REACH *Firefighting equipment does not have easy access to this farm fire.*

Hawaii *A helicopter drops fire accelerant to help a fire burn itself out. Below: A bulldozer is used to create a firebreak to check a forest fire.*

the fire, sometimes tens of miles away, using chainsaws or explosives to fell trees and bulldozers to clear the ground. This leaves a wide swath of open soil with no combustible material – the wider the firebreak the better. Sometimes they deliberately start another fire ahead of the conflagration, which burns up any fuel lying around and creates a wide area of scorched ground that cannot be burned again. Firebreaks are effective against surface fires, but sparks and burning embers from crown fires can be blown over the treated ground, landing a long way away. The key to successful firefighting lies in a good knowledge of weather conditions and in reliable predictions about the path and behaviour of the fire.

Knowledge is needed when it comes to fire prevention, too. The realisation, since the 1970s, that it was necessary to allow controlled fires in the national parks of North America, rather than to suppress all wild fires, entails a delicate balance. The land is now so heavily populated that it is impossible to achieve this balance without posing threats to settlement and civilisation. It has been noticed that wild fires have become hotter, more frequent and more devastating over the past few decades.

Fires do not advance along a gradual front. Drifting sparks and blown embers fall and set off fires – so-called 'spot fires' – at random points downwind of the main fire. This unpredictability is leading American fire specialists to produce computer programmes that will tell firefighters where the fire is likely to break out next. One programme, Farsight, takes into account the land's contours, the prevailing wind speed and direction, and the type of vegetation. Its first practical use in Yosemite in the summer of 1994 looked promising, and such programmes could help fire managers in the future to decide whether to tackle a blaze or to let it burn itself out.

Local Heroes *Firefighters work to contain a group of burning trees in a public park in Sydney, Australia.*

Water Bombs *A specially adapted helicopter drops water on a fire in New Zealand.*

Unpredictable Path *During a bush fire in Kenya, fires erupt in all directions as the flames are fanned by the wind.*

THE EXTERNAL THREAT

Our planet was built up from the impacts of millions upon millions of pieces of matter floating in space, all fusing together under the influence of gravity to form a solid body. However, the forces that created Earth could also destroy it.

It was to be the last fine day for a long time. Birds and other flying creatures were wheeling around the blue sky. Animals patrolled the lush forests of oak, birch and magnolia, with the occasional stands of conifer and a low growth of cycads and ferns. But few of these animals would be familiar to us, for this day marked the end of the Cretaceous period – the end of the age of the dinosaurs.

The event that proved the undoing of these animals came from something totally unexpected. For weeks, a piece of rock, maybe 6 miles (10 km) in diameter, had been circling in towards the Earth. An asteroid, like thousands of others in the Solar System, had drifted close to the Earth and been caught up in the Earth's gravitational field. By the time it had almost reached the centre of its inward spiral, not a single living creature on the ground knew anything of its existence – until a light appeared in the sky. This grew quickly into a dazzling fireball that streaked down at an angle, bulleting through the atmosphere at 62 000 mph (100 000 km/h) and disappearing behind the southern horizon. Suddenly there was a flash of light. The animals looked up, alarmed, but they could not comprehend what had happened. A jolt, as if from the heart of the Earth itself, cracked the ground surface, causing the great beasts to stagger and the huge trees to shift their roots in the soil. As the after image of the flash cleared, smoke billowed up away to the south, where forests exposed to the initial searing blast of heat and light had ignited instantaneously. Beyond this, the animals saw a glowing mushroom cloud swirl up into the sky far away, dimmed only by the haze of smoke and the distance. The cloud darkened as it towered upwards, flattening out as it reached through the various layers of the atmosphere.

As the shock wave of the impact finally approached at last, a noise rent the eardrums of the reptiles. The blast engulfed everything, wrenching trees from the ground, ripping away loose soil and rocks, and toppling the vast bodies of 5 ton dinosaurs. The air was a chaos of leaves and dust, none of which had time to settle before a backdraught sucked the debris and the bodies of stunned and dead animals back towards the site of the impact.

Then there was darkness, even as the flying debris from the shock wave and the thickest dust settled, for the spreading mushroom cloud had shrouded the Sun. Far away, towards the sea, this scene of complete destruction was followed by further devastation. The haze of settling dust was suddenly pushed aside by another thrust of air. A deep rumbling noise crescendoed to a climax, as what at first glance appeared to be a vast bank of approaching cloud proved to be a foaming cliff of water – a tsunami almost 1/2 mile (0.8 km) high. The nearby sea emptied itself across the surrounding lowlands, bringing with it the rocks and sediments wrenched up from the seabed – and instant oblivion to all life in the area. The shock waves spread through the Earth, jolting and juddering the entire world. Off the coast of Africa, loosely consolidated sediments on the continental shelf were shaken free and tumbled to the ocean depths, triggering vast tsunamis that engulfed animal life on the continent. The crust of the

SHOOTING STAR *A meteor trail appears when a dust fragment scorches across the evening sky and burns up.*

BARREN PIT *The impact of a meteor hitting the Earth causes a deep depression in the surface, as at Meteor Crater, Arizona.*

Earth in India, already weakened by plate-tectonic forces, split and basaltic lava erupted in a series of volcanoes that continued to pour out lava for centuries, eventually building up igneous masses that covered half of the subcontinent. Somewhere in the region that would one day hold the Caribbean Sea, there is a crater measuring 125 miles (200 km) in diameter.

In North America, the debris hurled into the atmosphere by the explosion began to fall. Huge boulders crashed down, fine pebbles pattered on the wrecked landscape, and finer material settled as a blanket. The asteroid had punched through the thin crust of the Earth, exposing the hot mantle, and had disintegrated with the impact. All its material, along with splinters of the Earth's crust and underlying mantle, spread throughout the atmosphere.

For weeks, the dark overcast sky was lit from beneath by the glow of forest fires. The leaves of the surviving plants were choked with soot and falling dust, unable to function in the dark.

END OF AN ERA *The worldwide destruction that took place 65 million years ago may have followed a meteorite impact.*

Clouds of moisture reflected sunlight back into space, and any sunlight that did filter through was absorbed by the black clouds of dust.

REGENERATION *Ferns would have survived the devastation of a meteorite impact.*

After a few months, the pall began to thin, as the bulk of the dust settled from the sky and sunlight began to peep through again. The water vapour that had been suddenly blasted into the atmosphere now fell as torrential rain, which flooded the plains and turned them into acidic bogs. The first plants to regenerate were those that could withstand these new conditions. Ferns sprouted from spores that had lain dormant through the worst of the chaos. Coniferous trees that could tolerate peat bogs germinated from seeds in the shattered cones that lay in the ash. Gradually, greenery returned to the land as the climates warmed with the return of the Sun. Small animals emerged from hibernation, as tiny mammals started to share the world with the smallest of the reptiles to survive.

AN ANCIENT DISASTER

No one can be sure that the Cretaceous period ended exactly like this, although the dinosaurs did die out some 65 million years ago, and geological evidence suggests that the previous scenario is very plausible. And it could happen again. The possibility of the Earth being hit by a lump of matter from space was not taken very seriously until a few years ago, except by science-fiction writers and film-makers. Then, in the early 1970s, the American geophysicists Luis and Walter Alvarez were studying rocks in Italy that formed the boundary between the Cretaceous period, when the dinosaurs were at their strongest, and the succeeding Tertiary period when the mammals took over. They were actually looking at how the natural magnetism of these rocks had changed through time, but they found something that sent shudders through the scientific world. In the midst of the limestone, there was a thin bed of clay, no more than 1 in (2.5 cm) or so thick, that coincided with the end of the Cretaceous period. This clay was totally devoid of fossils and was particularly rich in the element iridium, which is quite rare at the Earth's surface but is more common in meteorites. When they announced their findings in 1980, they hypothesised that the extinction of the dinosaurs, and 70 per cent of the remaining animal life, had been caused by a meteorite striking the Earth.

Scientists were sceptical at first, but subsequent work done by other researchers found traces of iridium dating from this time in many other parts of the world – Denmark, New Mexico and the sea floor off South Africa. In addition to this, the rocks contained fragments of the mineral quartz that bear the telltale marks of having been shattered by a vast explosion, and spherules of mineral that had been melted and solidified as flying drops. Researchers have also

THE EVIDENCE *An iridium-rich layer of clay between the Cretaceous and Tertiary periods is found in New Mexico (below) and in Italy (bottom).*

THE ULTIMATE NATURAL DISASTER

The Sun is expanding all the time. It may now be up to 30 per cent brighter than when it formed 4.5 billion years ago, and in 1.1 billion years' time it will have grown by a further 10 per cent. Some 3.5 billion years from now, the Sun will be 40 per cent hotter than it is today. The oceans will boil away, and life on Earth will be extinguished.

At some time thereafter – scientists disagree as to when – the Sun will balloon out into a burning red giant, 170 times its present diameter, and engulf the orbit of Mercury. In the process, its core will contract and it will lose about half its mass, and the planets' orbits will change. The Earth will drift outwards to about the orbit of Mars, but the heat from the expanding Sun will be so great that our planet's surface will be heated to 1873°C (3400°F) – the temperature at which rocks melt.

Finally, the Sun will collapse to a white dwarf and wink out, and the Earth will drift for ever as a lifeless cinder in the darkness of space.

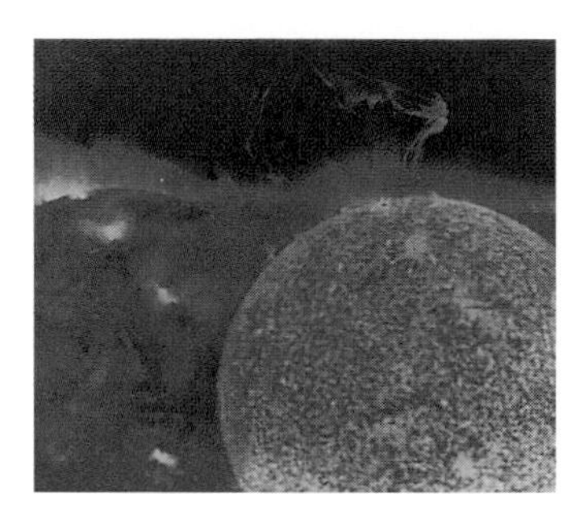

SOURCE OF LIFE *The energy from our life-giving Sun will not last forever.*

discovered deposits of soot from that time, which might have come from extensive forest fires. Furthermore, there are deposits of rock in the southern United States that may represent parts of the ocean floor that had been ripped up and redeposited on land by the action of tsunamis.

The works of Alvarez, father and son, have spurred astronomers into scanning the skies to find objects that might fall under the influence of the Earth's gravitational field and be drawn in. Since the 1980s, computers have helped with this research. Scientists are looking for 'Earth-crossing asteroids' with orbits that cut across the orbit of the Earth. Recent research has identified 100 such rocky bodies in the Solar System that have a diameter greater than 1/2 mile (1 km) – a figure that probably represents about 5 per cent of all existing asteroids. There are probably about 100 000 that have a diameter greater than 305 ft (100 m), and 150 million with a diameter greater than 30 1/2 ft (10 m). Any one of these could do untold damage if they struck the Earth.

THE MAKINGS OF THE EARTH

Some scientists believe that the Earth was formed by just such a process of meteorite strikes. The Solar System was created some 4.5 billion years ago, from a cold cloud of dust and gas that, as it started to spin, began to clump together at the centre. As the particles collided, setting off a chain of reactions, this central mass heated up to become the ancestral Sun. The rest of the material flattened out into a spinning disc, forming eddies of leftover material that began to accumulate in the form of planets. The Earth was one of these. Not all the material was gathered up as planets, however. A great deal was left over to form asteroids, minor planets and smaller bodies and lumps of rock. Throughout their histories, the Earth and all the other planets have been bombarded by material left over from their own creation.

There is yet another source of extraterrestrial material. Lumps of ice, with pieces of mineral scattered through them, swing through the Solar System on elliptical orbits, sometimes dropping towards the Sun and the inner planets. These ice masses are known as comets and may have their origin in a cloud of material way out beyond the limits of the Solar System – called the Oort cloud after the Dutch astronomer Jan Oort, who postulated its existence in 1950. There may be up to 10 trillion comets in this cloud, and every now and then the gravitational forces of our Sun and of other stars may disturb its equilibrium so much that a comet breaks away and drops down into the Solar System. As the comet passes close to the Sun, its ice begins to evaporate, producing a kind of atmosphere called a coma around its central solid part, which is known as the nucleus. The solar radiation blasts the looser material away into space, and this manifests itself as a tail trailing away from the Sun. There may be two tails – one of gas that forms a straight line, and one of solid particles that lags behind as the comet moves and is therefore seen as a curve. The dust may erupt at a rate of 10 tons per second, and the gas at twice that rate. When the Earth passes through the dust tail, the particles fall into the atmosphere and burn up in streaks of light called meteors.

The nucleus of the comet itself may even be caught in the planet's gravitational field and be pulled into a collision. This was witnessed in 1994 when a newly discovered comet, Comet Shoemaker-Levi 9, collided with the planet Jupiter. The comet's original orbit had probably been disturbed by the gravitational field of Jupiter as long ago as

THE OORT CLOUD *Around the outside of the Solar System is a cloud of material (right) left over from its creation. Above right: Burning dust shed by a comet can be seen against the movement of the stars.*

SHOOTING STARS *Halley's comet was visible from Earth in 1986.*

DIRECT HIT *In 1994 Jupiter was the target of a massive comet strike.*

1929 and from then, its ultimate collision with the planet would have been inevitable. As it neared the surface of Jupiter, the gravitational force became greater and greater until, sometime in 1994, it reached a stage where the side facing the planet was receiving a significantly greater pull than the side facing away. As a result, the comet was pulled to pieces, and it was the elongated string of about 20 comet fragments that astronomers first noticed. Analysing their trajectory, the astronomers deduced that this was the comet that had only been discovered 16 months previously, and that it had abandoned its original orbit of the Sun and was spiralling towards Jupiter. The successive impacts in July 1994 produced vast explosions, generating temperatures of 20 000°C (36 032°F) in the planet's atmosphere, blasting fireballs thousands of miles into space, and leaving marks on the face of the planet as the cometary material reacted with the dense atmospheric gases.

Such explosions and a similar atmospheric reaction would be expected if a comet collided with the Earth. In fact, we are fortunate in having Jupiter as a large gravitational mass in the outer Solar System. Eugene Shoemaker (one of the discoverers of the comet Shoemaker-Levi 9) believes that Jupiter's presence draws away many comets before they reach the inner Solar System. Without it there would be a hundred times as many comets passing close to us, and they would strike the Earth with a frequency of something like one every 100 years.

Some scientists think that the water in the Earth's oceans was derived from the ice of comets that collided with the Earth shortly after its formation. And it is possible that the demise of the dinosaurs at the end of the Cretaceous period was caused not by a meteorite, but by a cometary collision. This theory would overcome the initial opposition to the idea – the fact that when the meteorite theory was first proposed there was no sign of any meteorite crater. Being made of much lighter material than a meteorite, a comet would not have left a crater, but would have exploded in the atmosphere. It would have taken a series of comets to produce the environmental damage that seems to have taken place at the end of the Cretaceous period.

FIRE OVER SIBERIA

A cometary impact may, indeed, have occurred in recent history, although the event was almost totally ignored at the time. Around dawn on June 30, 1908, the skies

ASTRONOMICAL NAMES

Meteoroid is the general term referring to rocky matter in the Solar System, while **meteor** describes the phenomenon – the glowing head and the tail – seen when a meteoroid burns up in the Earth's atmosphere. Most meteors are formed from dust particles from the tails of comets. **Meteorite** is the name given to a meteoroid that has entered the atmosphere as a meteor and is found on the ground as a lump of rock. **Asteroids** or **minor planets** are rocky bodies that orbit the Sun between the orbits of Mars and Jupiter. Some have elliptical orbits that may bring them close to the orbit of the Earth.

TUNGUSKA, ALTERNATIVE VIEWS

The mystery of the Tunguska Event is complicated by the fact that it was not investigated for almost 20 years afterwards. No crater was found, nor any meteoritic fragments, and this has led to a great deal of speculation, informed or otherwise, about what happened.

Those who believe in unidentified flying objects and alien space vehicles suggest that one of these may have crashed in the area. If the nuclear power plant of such a vehicle had exploded, it would have produced precisely the effects observed at Tunguska.

Another theory is based on the fact that, at about this time, the Croatian-born American electrical pioneer, Nikola Tesla, was experimenting with a method of transmitting electricity across long distances without the use of cables. This theory's supporters suggest that, in June 1908, Tesla projected an electrical charge towards the North Pole from his laboratory on Long Island in an attempt to induce a display of the aurora borealis – the northern lights – to coincide with an American North Pole expedition. The charge, they claim, overshot its target and caused the devastation in Tunguska. The three sites – Long Island, the North Pole and the Tunguska River – lie on the same bearing.

FLATTENED *The devastation in the Tunguska area in 1908 may have been caused by a comet.*

THE CULPRIT? *Nikola Tesla lectures to scientists in Paris in 1892.*

above the basin of the Yenisei River in Siberia erupted in a white-hot fireball. The explosion was equivalent to that of a 15 megaton hydrogen bomb (or to 15 million tons of high explosive; by comparison, the bomb dropped on Nagasaki in 1945 was equivalent to 20 000 tons of high explosive). Since this was a sparsely populated region, no casualties were reported, but a number of meteorological anomalies were noticed in various parts of the Northern Hemisphere. British meteorologists observed strange fluctuations in atmospheric pressure, and some areas of Russia experienced particularly light nights, as sunlight was dissipated in high banks of dust cloud. It took a long time – 19 years, a world war and a revolution, to be precise – before a scientific expedition was dispatched to the area under the leadership of Leonid Kulik, the curator of the meteorite department in the Mineralogical Museum of Leningrad (now St Petersburg).

Kulik and his party travelled to the site by train, reindeer sledge and, finally, on snowshoes. As they approached the Tunguska River, where the event was supposed to have taken place, they began to notice burned patches in the forest, covered with 19 years' growth of vegetation. Then, at the top of a hill, they saw before them a scene of utter devastation. Trees had fallen – all pushed over in the same direction, like the pickets of a fence. Inward-facing areas on top of the surrounding hills had been swept clear of any vegetation. Something immense had obviously happened here, but Kulik had no time to investigate further as the spring thaw was spreading and making the ground waterlogged and impassable.

He returned later, and was able to conduct a thorough survey. An area 40 miles (64 km) in diameter had been flattened, with trees on the ground, all facing away from the centre. A deposit of charred soil and vegetation was evidence of the intensity of the heat. Trees in sheltered hollows were still upright but they had been incinerated. There was no sign of meteorite fragments and no vestige of a crater. The absence of such signs was explained initially by the fact that the area lay at the edge of the permafrost – a region in which it is so cold that the deep soil remains frozen all year. Only the surface layers thaw in the spring, and the result is a boggy mass of swamp and lake that cannot drain away. A meteorite crater melted into the permafrost would flood, freeze and fill up within a few winters, burying any scattered pieces of meteoritic rock as well. The current thinking is that the explosion was caused by a comet,

EXPLOSION FOOTPRINT *Today the area of the Tunguska impact is densely forested.*

EYEWITNESSES?

In New South Wales, Australia, is a series of 13 craters formed by an iron meteorite some 5000 years ago. Aboriginal legends of fire in the sky and flying devil rock may refer to this incident as seen by their ancestors. In South Island, New Zealand, serious environmental disruption occurred about 800 years ago. Maori legends tell of a falling of the skies and upheavals of the earth, and this may be a reference to an observed meteorite impact.

IMPACT CRATERS *Meteorite craters, as seen on Jupiter's moon Callisto (top), Mercury (middle), and our own moon (above), are common in the Solar System.*

with a solid portion 100-300 ft (30-91 m) wide, which entered the Earth's atmosphere at a speed of 30 000 mph (48 000 km/h) and exploded at an altitude of about 5 miles (8 km), leaving no crater on the ground.

THE SCARS

The surface of the Moon is covered with craters, dating from very early in the history of the Solar System. Direct sampling of material from these craters by the Apollo expeditions of the 1960s and 1970s showed them to be of meteoritic origin, not volcanic as had previously been suspected. Mercury, too, is scarred by meteorite craters, as are the moons of the outer planets. Jupiter's moon, Callisto, is regarded as the most cratered object in the whole of known space – it would not be possible to add another crater to Callisto without obliterating existing craters. The Earth itself cannot have avoided being hit and cratered by giant meteorites at some time in its history. But if that is so, where are these craters? Most of them would have been formed in the first billion years or so of the life of the Solar System and would have been wiped away long ago, by the constant forces of erosion and the movements of the Earth's plates. Nevertheless, there have been large meteorite strikes in more recent times, and the scars are still visible.

Perhaps the most famous impact site lies in Arizona. Known during the 19th century under a number of names, including Coon Butte, Franklin Hole and Crater Mountain, it consists of a vast circular hole in the desert, 4000 ft (1200 m) in diameter and about 570 ft (175 m) deep. Its raised rim lies about 150 ft (45 m) above the surrounding desert surface. Between 1890 and the 1920s, several expeditions were sent to

EARTH'S SCARS *Terrestrial impact sites include Meteor Crater in Arizona (below) and Wolfe Creek Crater in Western Australia (bottom).*

the site, which confirmed its true nature and gave it a new name, Meteor Crater. At first, the crater was thought to mark the point where a meteorite had penetrated the Earth and buried itself under the ground. However, geophysical investigation and drilling in the 1920s – funded by companies hopeful of discovering a vast lump of valuable iron – proved that there was no buried meteorite beneath the structure. It became plain that the crater was the result of an explosion, during which the meteorite had disintegrated on impact, pulverising the rocks in the immediate vicinity and forcing up the sediments to form the crater rim. Subsequent investigations revealed large quantities of meteoritic material scattered across the surrounding desert.

The age of Meteor Crater has been estimated at between 5000 and 50 000 years – very recent in geological terms. Even so, the relentless forces of erosion have been working upon it. The flanks of its bowl shape are cut by gullies from the infrequent but fierce desert rainstorms; and loose material has slumped and settled on the bottom. If these processes can affect such a big structure in such a short time then it may be difficult to discover more ancient meteor craters. Yet as many as 200 older structures have been identified.

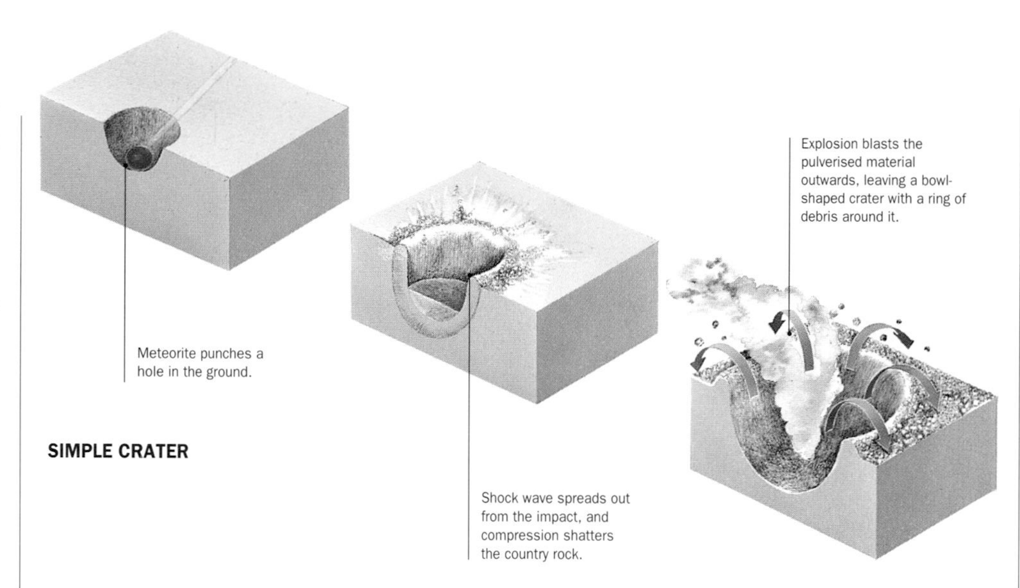

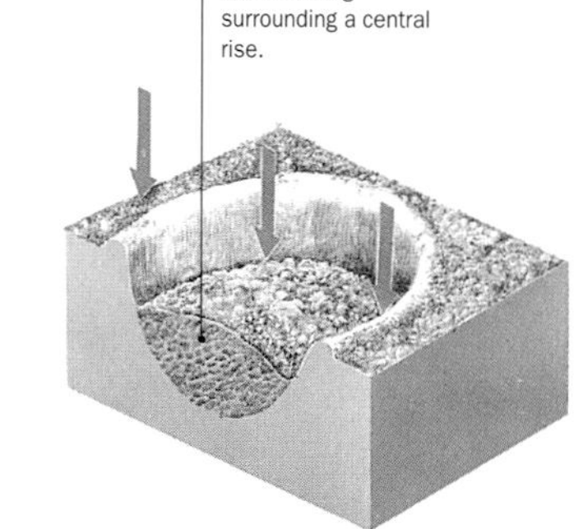

IMPACT *A meteorite blasts out a crater and piles up debris round it (above). Large impacts can disturb the bedrock, creating a complex crater (right).*

They come in two general formations. The smaller craters, such as Meteor Crater in Arizona, are known as 'simple craters' and consist of a bowl-shaped depression. The bottom may have a layer of molten and solidified rock generated by the impact, and the depression is partially filled with material that has been blasted out and fallen back in, to be covered by later sedimentary material. The larger craters, such as the Manicougan Crater in Quebec, are the result of such a massive impact that the whole geological structure of the area is affected, and a 'complex crater' results. The rock is drawn inwards and upwards, forming a rise in the centre of a relatively shallow, ring-shaped depression.

Some suspected meteorite craters are very large indeed – particularly if you consider that were the world's entire nuclear arsenal to be detonated all at once in one place it would produce a crater 6 miles (10 km) in

AN EXTERNAL VIEW OF THE EARTH'S INTERIOR

There are two main types of meteorite. The stony type has a surface blackened by the heat of its passage through the atmosphere. The iron type is usually pockmarked with 'thumb-prints', the effect of melting by friction.

The Earth is built up of layers. An iron core is surrounded by a stony mantle, which is covered by a lighter, stony crust. These layers are thought to reflect the composition of the original cloud of gas and dust that formed the planet. As the materials accumulated, the heaviest – iron – sank towards the centre and the lighter remained on the outside. The relative numbers of heavy iron meteorites and lighter stony meteorites are similar to the proportions of iron and stony material in the structure of Earth, suggesting that they all come from the same material.

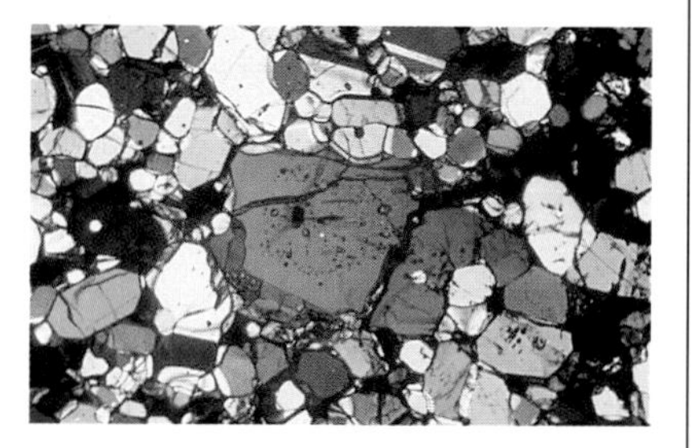

CLOSE VIEW *Microscopy with polarised light allows meteorite minerals to be identified.*

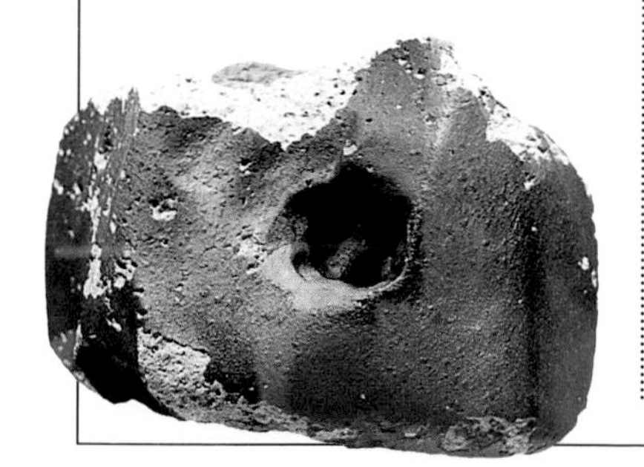

OLD STONE *This meteorite from Antarctica dates back 1300 million years.*

IDENTIFIED AND POSSIBLE METEORITE CRATERS

Crater	Location	Diameter	Age
Wolf Creek Crater	*Western Australia*	*1/2 mile (0.88 km)*	*A few thousand years*
New Quebec Crater	*Northern Quebec, Canada*	*2.2 miles (3.5 km)*	*A few thousand years*
Meteor Crater	*Arizona, USA*	*3/4 mile (1.2 km)*	*5 000-50 000 years*
Ries Crater	*Germany*	*15 miles (24 km)*	*14.8 million years*
Praha Basin	*Czech Republic*	*200 miles (320 km)*	*15 million years*
Popigai Crater	*Siberia*	*65 miles (105 km)*	*30-65 million years*
Chicxulub Structure	*Yucatan*	*110 miles (175 km)*	*65 million years*
Mjlnir Structure	*Barents Sea*	*25 miles (40 km)*	*125-161 million years*
Doulun Crater	*China-Mongolia border*	*45 miles (72 km)*	*165 million years*
Manicougan Crater	*Quebec, Canada*	*62 miles (100 km)*	*210 million years*
Lake Huron Crater	*Canada-USA border*	*30 miles (48 km)*	*500 million years*
Holleford Impact Structure	*Ontario, Canada*	*1.4 miles (2.3 km)*	*500-1000 million years*
Sudbury Complex	*Ontario, Canada*	*124 miles (200 km)*	*1.8 billion years*

diameter. A fraction of this would be enough to generate the 'nuclear winter' so dreaded in the 1980s. The damage inflicted by the event that produced the 200 mile (320 km) diameter Praha Basin in the Czech Republic 15 million years ago must have been colossal.

DEBRIS RING ***Henbury meteorite crater, Northern Territory, Australia, forms a distinctive landscape feature.***

How likely is such a disaster to happen during the tenure of human civilisation on Earth? The big meteorite strike that brought the dinosaurs to an end involved volcanic eruptions, earthquakes, tsunamis, landslides, avalanches, hurricane winds, fire, flood and climatic disruption – in fact, all the catastrophes covered in this book. Is this likely to happen again?

Despite all the research, we still do not know how many dangerous objects are floating in the Solar System. On March 22, 1989, an asteroid 1/2 mile (1 km) in diameter passed within 430 000 miles (690 000 km) of the Earth – a near miss in astronomical terms. But we knew nothing about it until it had passed. On October 30, 1937, Hermes, an asteroid twice as big, had passed almost as close.

A meeting of scientists in the University of Chicago in 1986 argued that the world's defence budget could be diverted into preparation for natural disasters. Perhaps it would be sensible to create international networks of seismic sensors, to establish international standards for earthquake-proof building, and to keep two years' supply of food in storage out of world surpluses. Furthermore, Cold War technology could provide the basis for the detection and aversion of meteorite strikes.

A number of authorities have come up with schemes for averting any impact disaster. In the days of the Cold War, the United States experimented with a system called the Strategic Defence Initiative – sometimes nicknamed 'Star Wars' – which was to have comprised satellite-based laser weapons designed to destroy ballistic missiles fired at the North American continent. It has been suggested that the technology developed for this could be turned against any incoming meteoroid, destroying or deflecting it as it approached. However, we are far from having the ability to deploy such a system, and we do not yet even have one that can detect a foreign body as it approaches Earth – as illustrated by the fact that the 1989 asteroid was not noticed until it had been and gone.

It has been calculated that the frequency of something 300 ft (91 m) in diameter plunging through the atmosphere is once every 50 to 100 years. Perhaps once every 100 000 to 200 000 years, there will be an impact big enough to cause a global catastrophe. A meteorite 1 1/4 miles (2 km) in diameter would wipe out a quarter of the world's population. Averaged out over time, this makes a meteorite strike a bigger threat to human life than air travel.

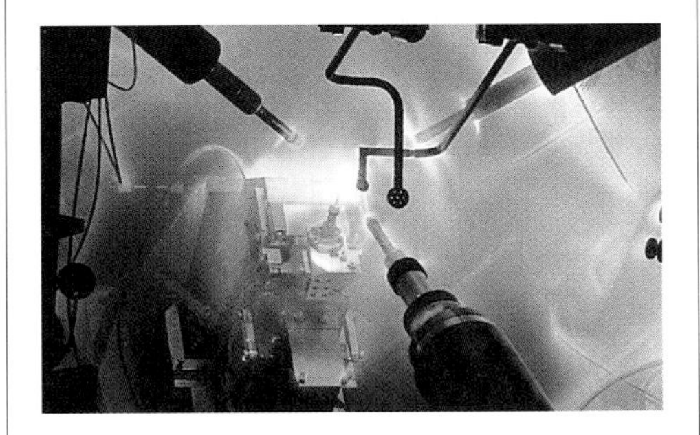

STAR WARS ***Laser weapons developed for nuclear warfare may someday be deployed against incoming meteorites.***

INDEX

Page numbers in *italic* denote captions or illustrations.

PICTURE CREDITS

Abbreviations:
T = top; C = centre; B = bottom;
R = right; L = left
BCL= Bruce Coleman Ltd
KP= Katz Pictures
RFL= Rex Features Ltd
SPL= Science Photo Library
SAL= Sienna Artworks Ltd, London

3 Frank Spooner Pictures. **6** Hulton Getty, CL; BCL, CR; BCL/Peter Davey, BR. **7** Tony Stone Images/Barbara Filet, CL; Popperfoto/Mike Nelson, BL; Holt Studios International/Nigel Cattlin. **8** The Mansell Collection, TL; SPL/Noah Poritz, C; NHPA/Hellio & Van Ingen, BR. **9** FLPA/Silvestris, TL; Sygma/N. Hashimoto, B. **10** Robert Harding Picture Library/Vulcain. **11** Still Pictures Environmental Agency/Gerard & Margi Moss, TR; Popperfoto, BR. **12** Zefa. **13** KP, TL; Telegraph Colour Library/William Waterfall, BR. **14** Popperfoto. **15** BCL/Werner Stoy. **16** SPL/David Weintraub, TL; SAL/Ed Stuart. **17** SAL/Ron Hayward. **18** G.R. Roberts, CL; Ardea London Ltd/Parer-Cook, BL; SAL/Leslie Smith. **19** BCL/Raimund Cramm GDT, TL; Ardea London Ltd/François Gohier, BL; GeoScience Features Picture Library, CR; SAL/Leslie Smith. **20** Sygma, CL; BCL/Keith Gunnar, BL; SAL/Ron Hayward. **21** RFL, TL; Ardea London Ltd/François Gohier, TR, BR. **22** Frank Spooner Pictures, TL; Colorific/Michael Yamashita, BL. **22-23** Ardea London Ltd/Adrian Warren. **24** KP/Robert Seger, BL; OSF/Dieter & Mary Plage, TR. **25** OSF, TL; Frank Spooner Pictures, CR; RFL, BR. **26** RFL. **26-27** KP/Oxley/Compix. **27** Frank Spooner Pictures, TC; RFL, C. **28** RFL, TL; SPL/NRSC Ltd, BL; GeoScience Features Picture Library, BR. **29** KP/Van Cappellen, CL, TR; RFL, BL. **30** Ardea London Ltd CL; SPL/Simon Fraser, BL; FLPA/Mark Newman, TL; Comstock/Mike Andrews, BR. **31** GeoScience Features Picture Library, CL; Colorific/James Mason, TR; SPL/Laboratory for Atmospheres/NASA/GSFC, BL; BC; BR. **32** Frank Spooner Pictures, TL; SPL/NASA, TR; The Bridgeman Art Library/Tate Gallery, London, BR. **33** Ardea London Ltd/Adrian Warren. **34** Dr Robin Adams. **35** Sygma/Christian Simonpietri. **36** Dr Robin Adams. **37** Hulton Getty, BL; Dr Robin Adams, TR. **38** OSF/Terry Middleton, TR. **39** KP/George Hall, TL; Frank Spooner Pictures, TC, TR. **38-39** SAL/Leslie Smith. **39** SPL/Massonnet et al/CNES, BC. **40** Colin Woodman, TL; Corbis-Bettmann, BR. **41** SPL/David Parker, TL; KP, BL; Colin Woodman, CR. **42** SPL/David Parker, TC. **42-43** Frank Spooner Pictures. **43** Popperfoto, TR. **44** Popperfoto. **45** RFL, TL, BL; SAL/Ed Stuart. **46-47** RFL, TC; Sygma/Patrick Robert, BC. **47** Sygma/Patrick Robert, TR. **48-49** Colorific/Alon Reininger. **49** KP, TR; SAL/Ron Hayward. **50** SPL/David Parker. **51** KP. **52** NOAA/EDIS, C; US Army Corps of Engineers, BC. **53** FLPA. **54** OSF/Dieter & Mary Plage, TR; SAL/Ron Hayward. **55** Mary Evans Picture Library, TL; The Mansell Collection, BL. **56** Zefa, BL; SAL/Ron Hayward. **57** NOAA, TL, CL. **58** The Ronald Grant Archive, CL, TR, BR. **59** FLPA/S. McCutcheon, TL. **61** Chris Bonington Library, TL; NHPA/Eric Soder, BR. **62** SPL/NRSC Ltd. **63** BCL/Jeff Foott Productions. **64** GeoScience Features Picture Library, TR; SAL/Pavel Kostal. **65** Zefa, TL, TR. **66** Tony Waltham, TR, CR, BC; SAL/Pavel Kostal. **67** FLPA/S. McCutcheon, TR, BL; SAL/Pavel Kostal. **68** GeoScience Features Picture Library, TR; Tony Waltham, BL; SAL/Pavel Kostal. **69** OSF/Richard Packwood, B; SAL/Pavel Kostal. **70** Popperfoto, TR; The Environmental Picture Library, BL. **71** John Clear/Mountain Camera. **72** BCL/Steven Kaufman, TL; Mireille Vautier, BR. **73** Colorific/Carlos Humberto, CL, BL; Tony Waltham, BR. **74** Tony Waltham, TL, CR. FLPA/Silvestris, CL. **75** OSF/Stan Osolinski. **76** Keystone Press AG Archive, C, BC. **77** FLPA/Silvestris. **78** SPL/Claude Nuridsany & Marie Perennou, CL; Chris Bonington Library, BR; SAL/Pavel Kostal. **79** Keystone Press AG Archive, CL, TR; SAL/Pavel Kostal. **80** The Bridgeman Art Library/Pinacoteca Capitolina, Palazzo Conservatori, Rome, TR; FLPA/Roch, BL; John Cleare/Mountain Camera, BC. **81** Keystone Press AG Archive, TL; Patrick Cone Photography, CL; Landform Slides, BR. **82-83** Chris Bonington Library/Alan Hinkes. **83** SPL/W. Bacon, TR. **84** Keystone Press AG Archive, TL, CL, BL; Patrick Cone Photography, BC, TR. **85** OSF/Tony Martin, BL; Frank Spooner Pictures, TR; Mountain Camera/Leo LeBon, BR. **86** OSF/Doug Allan, BL. **87** SPL/NASA, TL; Zefa/A & J Verkaik, BR. **88** Corbis-Bettmann. **89** SPL/Sam Pierson. **91** SPL/NOAA, CL; Telegraph Colour Library/J. Zehrt, TR; SAL/Pavel Kostal. **92** Sygma, CR; Zefa, BR. **93** Magnum Photos/Misha Erwitt, TL, BL. **94** Tom Stack & Associates/Larry Lipsky, TL, BL. **95** Colorific/Boccon-Gibod/Black Star, TL; Magnum Photos/James Nachtwey, B. **96** FLPA, TR; SPL/Peter Menzel, BL; Tom Stack & Associates/NCAR, CR. **97** Popperfoto/AFP. **98** FLPA. **99** SPL/Fred K. Smith. **100** FLPA/David Hoadley, TL, TC; SAL/Pavel Kostal. **101** Tom Stack & Associates/Merrilee Thomas, TR. **102** FLPA/D. Kinzler, TL. **102-3** FLPA. **104** Tom Stack & Associates/Kenneth Langford, TR; FLPA/Robert Steinau, CR; FLPA/David Hoadley, BR. **105** Tom Stack & Associates/Brian Parker. **106-7** Zefa/A & J Verkaik. **108** Telegraph Colour Library/Stan Osolinski, CL; Telegraph Colour Library/Planet Earth Pictures/Paolo Fanciulli, BL; FLPA, CR. **109** FLPA. **110** Royal Collection Enterprises Ltd, C; Topham Picturepoint, BC. **111** OSF/Lloyd Nielson. **112** Topham Picturepoint, TL; Popperfoto, TR; The Bridgeman Art Library/Chiesa di Santa Maria Novella, Florence/K & B News Foto, Florence; Scala/Santa Croce, Florence, BL, BC. **113** KP/Tavernier, TL; Camera Press, CR, BR. **114** SPL/Earth Satellite Corp, TC, TR; KP, B. **115** The Environmental Picture Library/Rob Visser, TC; Comstock/Cameron Davidson, TR; DRK Photo/Wayne Lynch, BL. **116** Patrick Cone Photography. **117** NHPA/Roger Tidman, CL; NHPA/David Woodfall, TR; Popperfoto, BR. **118** BCL/Paul van Gaalen, BL; Magnum Photos, CR; SAL/Ron Hayward. **119** Auscape International/Wayne Lawler, CL; G.R. Roberts, BL; DRK Photo/Stephen Krasemann, TR; SPL/Dr Robert Spicer, CR. **120** Popperfoto, TL, C; Topham Picturepoint, BR. **121** RFL, TL; Auscape International/J.M. La Roque, BR. **122-3** Auscape International/J.M. La Roque. **123** Comstock/Georg Gerster, TR; Nik Wheeler, CR. **124** BCL/Keith Gunnar, BL; BIOS/Cyril Ruoso, TR; KP/Agenzia Contrasto, CR. **125** Tom Stack & Associates/Doug Sokell. **126** Robert Harding Picture Library, C; Comstock/Georg Gerster, BC. **127** Planet Earth Pictures/Thomas Dressler. **128** The Mansell Collection, TL; SAL/Ron Hayward. **129** Magnum Photos/Bruno Barbey, BC; OSF/Marty Cordano, BR. **130** Topham Picturepoint, CL, BL. **131** DRK Photo/Michael Fogden, T; Magnum Photos/Steve McCurry, BR. **132** The Environmental Picture Library/C. Jones, T; SPL/NASA, BR. **133** Topham Picturepoint, TL; Magnum Photos/Pinkhassor, CR; KP, BC. **134** BCL/Frans Lanting, TL; Planet Earth Pictures/Hans Christian Heap, BR; Comstock/Julian Nieman, CR. **135** SPL/Betty Milon, TL; BIOS/Michael Gunther, BR. **136** BIOS. **137** DRK Photo/Belinda Wright. **138** OSF/Richard Davies, CL; **138-9** RFL. **139** Auscape International/Reg Morrison, TC; Auscape International/Jean-Paul Ferrero, TR. **140** The Mansell Collection, BL; RFL, TC; BCL/Jeff Foott, TR. **141** Auscape International/D. Parer & E. Parer-Cook, BL; OSF/Marty Cordano, BC; DRK Photo/Stephen Krasemann, BR. **142** Ardea London Ltd/François Gohier, TL; OSF/John Gerlach, TC; DRK Photo/Stephen Krasemann, BR. **143** Novosti, TC; Patrick Cone Photography, BR; **144** G.R. Roberts, TL, BL; Planet Earth Pictures/David E. Rowley, CL; Popperfoto, TR. **145** OSF/Steve Turner. **146** SPL/David Nunuk. **147** BCL/Steve Kaufman. **148-9** Kevin Jones Associates. **149** DRK Photo/D. Cavagnaro, TC; SPL/Dr Robert Spicer, CR; SPL/Prof. Walter Alvarez, BL. **150** SPL/NASA, TC; SPL/Pekka Parvianen, CR; SPL/Dr S. Shostak, BR. **151** SPL/Magrath Photography/Nielsen, TR; SPL/NASA, TL. **152** Novosti, TL; Mary Evans Picture Library, TR; Comstock/Georg Gerster, BL. **153** SPL/NASA, TC, TL, CL; SPL/François Gohier, CR; Tom Stack & Associates, BR. **154** SPL/NASA, BL; SPL/Ian Steele & Ian Hutcheon, C; SAL/Ron Hayward. **155** Planet Earth Pictures/Hans Christian Heap, CL; SPL/Alexander Tsiaras, CR.

FRONT COVER: Zefa/A & J Verkaik; Icelandic Photos, C.